Competition Algebra

Xing Zhou

Math for Gifted Students

http://www.mathallstar.org

Copyright © 2015 by Xing Zhou. All rights reserved.

No part of this book may be reproduced, distributed or transmitted in any form or by any means, including photocopying, scanning, or other electronic or mechanical methods, without written permission of the author.

To promote education and knowledge sharing, reuse of some contents of this book may be permitted, courtesy of the author, provided that: (1) the use is reasonable; (2) the source is properly quoted; (3) the user bears all responsibility, damage and consequence of such use. The author hereby explicitly disclaims any responsibility and liability; (4) the author is notified in advance; and (5) the author encourages, but does not enforce, the user to adopt similar policies towards any derived work based on such use.

Please visit the website http://www.mathallstar.org for more information or email contact@mathallstar.org for suggestions, comments, questions and all copyright related issues.

use your mobile device to scan this QR code for more resources including books, practice problems, online courses, and blog.

This book was produced using the LaTeX system.

Contents

1 **Introduction**   1

2 **Vieta Theorem**   3
   2.1 Solving Quadratic Equation 3
   2.2 Vieta's Theorem 4
   2.3 Frequently Used Techniques 6
     2.3.1 Polynomial Transformation 6
     2.3.2 Degree-Reducing 7
     2.3.3 Recursion 8
     2.3.4 Equation Transformation 9
     2.3.5 Equation Construction 10
     2.3.6 Properties of Roots 11
   2.4 Practice 12

3 **High Degree Equation**   19
   3.1 Introduction 19
   3.2 Factorization 19
   3.3 Rational Zero Theorem 20
   3.4 Substitution 22
   3.5 $n+1$ Roots 23
   3.6 Practice 25

4 **Non-Polynomial Equation**   29
   4.1 Absolute Value 29
   4.2 Irrational 32
   4.3 Fraction 33
   4.4 Floor and Ceiling 36
   4.5 Practice 38

5 **Sequence**   43
   5.1 Introduction 43
   5.2 Basic Sequence 44
     5.2.1 Arithmetic Sequence 44
     5.2.2 Geometric Sequence 47

CONTENTS

    5.3 Linear Recursion 48
          5.3.1 $a_{n+1} = pa_n + q$ 48
          5.3.2 $a_{n+2} + pa_{n+1} + qa_n = 0$ 50
    5.4 Special Sequences 54
    5.5 Practice 56

6 Function   61

    6.1 Domain and Range 61
    6.2 Function Properties 64
          6.2.1 Odd Function v.s. Even Function 64
          6.2.2 Periodic Function 65
          6.2.3 Monotonic Function 66
    6.3 Function Equation 69
    6.4 Practice 71

7 Solutions   75

    7.1 Introduction 76
    7.2 Vieta Theorem 77
    7.3 High Degree Equation 94
    7.4 Non-Polynomial Equation 103
    7.5 Sequence 113
    7.6 Function 126

Preface

Welcome to Math All Star© series!

Math All Star originates from a series of lectures given to a group of gifted middle school students with a love for mathematics and an interest in participating in competitions such as MathCounts, AMC, and AIME. These lectures aim to strengthen their problem-solving abilities and to introduce effective techniques that are not typically taught in the classroom.

As the popularity of Math All Star grew, the author began to upload lecture materials to create online courses, thereby providing students with the opportunity to progress at their own paces.

Since then, course materials have constantly been reviewed and updated to reflect student feedback and the observations made during lectures. Recent competition problems are also continuously analyzed and referenced to ensure the relevance of the contents. These course materials are the foundations of this Math All Star series.

Because competition math is a diversified subject that covers both a wide breadth and depth of topics, it is quite challenging to effectively cover all the material in one book that is appropriate for every interested student. Consequently, the author has decided to write a series of books, with each one focusing on a particular topic. Students are encouraged to pick and choose where to begin, depending on their individual skill levels and needs.

CONTENTS

In addition to these books, the Math All Star website provides extra practice problems and serves as a highly recommended supplemental learning resource.

If there are any questions, comments, or concerns, please visit the website or email `contact@mathallstar.org`.

Happy learning!

To visit the Math All Star website, scan this QR code or go directly to
http://www.mathallstar.org

Chapter 1

Introduction

Algebra is taught from elementary school to college and beyond. Algebraic problems present a significant portion in all math competitions including MathCounts, AMC, AIME, USAMO and so on. Therefore, solving competition level algebraic problems is a must-master skills for every contest contender.

Algebra includes a wide range of topics and techniques. Some of them may be related to advanced mathematical theorems and tools. Therefore, it is impossible to cover all of them in one book. However, middle school and high school level competitions usually do not require advanced mathematics. Instead, the emphasis is on the applications of basic algebraic skills in a flexible and effective way to solve complex problems. As a result, it is a wise strategy to thoroughly understand the most important topics and drill down into details of related solving techniques in order to improve one's skill and test performance.

This book covers three basic but important topics: equation, sequence and function. While these topics are all taught in schools, there are some competition specific techniques which deserve a systematic discussion.

Chapter 1: Introduction

Taking Vieta's theorem as an example. While polynomial transformation is a well known method to evaluate expressions such as $x_1^2 + x_2^2$, there are several other powerful techniques. They can be used to evaluate some complex expressions in a more efficient and less error-prone way. These expressions can have high power such as $x_1^7 + x_2^7$, or are asymmetric such as $5x_1^3 + 3x_2^5$. In fact, the latter asymmetric expression can present a challenge to many students who only know the polynomial transformation method. In addition to expression evaluation, Vieta's theorem can also be used to solve some seemingly unrelated problems. Such problems are among top hits in various math competitions.

Sequence is another good example. Most students understand the two basic types of sequences, namely, arithmetic and geometric. Though the vast majority of sequence related problems in math contests can be converted to these basic types, finding such conversion may be a demanding task which is usually not discussed in classrooms. Meanwhile, in order to become a strong competitor, one must also understand a few additional more complex sequences especially those defined recursively. They are beyond the scope of school textbooks, but are discussed in this book.

The goal of this book is to give an organized in-depth discussion on competition level techniques. Fully understanding these techniques will help students to quickly recognize and solve these types of problems. It will also lay down a solid foundation for them to solve other problems whose solutions require these algebraic techniques as critical stepping stones.

Chapter 2

Vieta Theorem

2.1 Solving Quadratic Equation

Solving quadratic equation is one of the most important basic skills in algebra. Given a quadratic equation

$$ax^2 + bx + c = 0 \qquad (a \neq 0)$$

its two roots are given by

$$x_{1,2} = \frac{-b \pm \sqrt{b^2 - 4ac}}{2a} \qquad (2.1)$$

Some problems can be solved without calculating their roots directly. They can be tackled by exploiting the relation between an equation's roots and its coefficients. Such relation is known as *Vieta's formula* which is also referred as *Vieta's theorem*. In this chapter, we will look into Vieta's theorem and its applications.

> The key to solve Vieta's theorem related problems is NOT to solve the roots directly.

Chapter 2: Vieta Theorem

2.2 Vieta's Theorem

Adding and multiplying the two roots in (2.1) lead to

$$x_1 + x_2 = -\frac{b}{a} \quad \text{and} \quad x_1 \cdot x_2 = \frac{c}{a}$$

These two relations hold regardless whether $x_{1,2}$ are real numbers or not. They are the quadratic form of Vieta's theorem.

> **Theorem 2.2.1 Vieta's Theorem (Quadratic)**
>
> Let x_1 and x_2 be the two roots of quadratic equation $ax^2 + bx + c = 0, (a \neq 0)$. Then
>
> $$x_1 + x_2 = -\frac{b}{a} \quad \text{and} \quad x_1 \cdot x_2 = \frac{c}{a}$$

In addition to utilizing (2.1), these two relations can also be proved in the following way.

Proof

Because x_1 and x_2 are the two roots of $ax^2 + bx + c = 0$, the following factorization must hold:

$$ax^2 + bx + c = a(x - x_1)(x - x_2) = ax^2 - a(x_1 + x_2)x + ax_1x_2$$

Comparing the coefficients of the two sides leads to

$$\begin{cases} b = -a(x_1 + x_2) \\ c = ax_1x_2 \end{cases} \implies \begin{cases} x_1 + x_2 = -\dfrac{b}{a} \\ x_1 x_2 = \dfrac{c}{a} \end{cases}$$

QED

This proof is the better for two reasons. Firstly, it does not require the root formula *(2.1)*. This is not only convenient but also important because root formulas for higher degree equations are usually unavailable. The second reason is that this proof can be applied to prove Vieta's theorem for higher degree equations as well.

> **Theorem 2.2.2 Vieta's Theorem (n-degree)**
>
> Let $x_1, x_2, \cdots, x_n$ be the roots of equation $a_0 x^n + a_1 x^{n-1} + \cdots + a_{n-1} x + a_n = 0$ where $a_0 \neq 0$. Then
>
> $$\begin{cases} x_1 + x_2 + \cdots + x_n = -\dfrac{a_1}{a_0} \\[1em] x_1 x_2 + x_1 x_3 + \cdots + x_1 x_n + x_2 x_3 + \cdots = +\dfrac{a_2}{a_0} \\[1em] x_1 x_2 x_3 + x_1 x_2 x_4 + \cdots + x_2 x_3 x_4 + \cdots = -\dfrac{a_3}{a_0} \\[1em] \cdots = \cdots \\[1em] x_1 x_2 \cdots x_n = (-1)^n \dfrac{a_n}{a_0} \end{cases}$$

Its proof can be done by comparing the coefficients between

$$a_0 x^n + a_1 x^{n-1} + \cdots + a_{n-1} x + a_n = 0$$

and

$$a_0 (x - x_1)(x - x_2) \cdots (x - x_n) = 0$$

Please note that the kth expression in *Theorem 2.2.2* is the sum of all the possible products of k roots. Therefore, the 1st expression has n terms, the 2nd expression has C_n^2 terms, the 3rd expression has C_n^3 terms, and so on.

> Do not miss any terms when employing the Vieta's theorem.

Chapter 2: Vieta Theorem

2.3 Frequently Used Techniques

Let's first consider the following example.

Example 2.3.1

Let x_1 and x_2 be the two roots of $x^2 + 2x - 3 = 0$, evaluate the following expressions without computing x_1 and x_2 directly.

$$(i) \quad x_1^2 + x_2^2 \qquad (ii) \quad \frac{1}{x_1} + \frac{1}{x_2}$$

Admittedly, this problem can be solved without using Vieta's theorem because it is simple enough to compute the two roots directly. Hence, all these target expressions can be evaluated easily. However, Vieta's theorem related problems in real tests will involve hard-to-solve equations almost certainly.

The purpose of presenting such a simple equation here is to facilitate easy result validation. Because solutions do not depend on the exact values of these two roots, whether the given equation is easy to solve or not is irrelevant to the demonstration of these techniques.

2.3.1 Polynomial Transformation

Polynomial transformation is the method of choice when solving Vieta's theorem related problems such as *Example 2.3.1*. The objective is to convert the target expression to a combination of $(x_1 + x_2)$ and $(x_1 \cdot x_2)$ whose values can be obtained by using Vieta's theorem.

Let's solve *Example 2.3.1*. By Vieta's theorem, $x_1 + x_2 = -2$ and $x_1 \cdot x_2 = -3$. Therefore

$$x_1^2 + x^2 = (x_1 + x_2)^2 - 2x_1 x_2 = (-2)^2 - 2 \times (-3) = \boxed{10}$$

and
$$\frac{1}{x_1} + \frac{1}{x_2} = \frac{x_1 + x_2}{x_1 \cdot x_2} = \frac{-2}{-3} = \boxed{\frac{2}{3}}$$

This is a perfectly appropriate technique to solve this problem. However, when the target expression becomes more complex or irregular, polynomial transformation may become tedious and thus prone to errors. For example, transforming $(x_1^3 + x_2^3)$ is still manageable but handling $(x_1^{10} + x_2^{10})$ can present a practical hassle.

2.3.2 Degree-Reducing

The previous technique utilizes general polynomial identities, but not problem specific information. For example, the identity $x_1^2 + x_2^2 = (x_1 + x_2)^2 - 2x_1 \cdot x_2$ always holds even if x_1 and x_2 are not the roots of the given equation. This implies that such a technique may not have fully utilized all the useful information.

The degree-reducing technique relies on the given equation in order to simplify the target expression. For example, the equation given in *Example 2.3.1* can be written as $x^2 = 3 - 2x$. Meanwhile, because x_1 and x_2 are two roots, it must hold that $x_1^2 = 3 - 2x_1$ and $x_2^2 = 3 - 2x_2$. Therefore,

$$x_1^2 + x_2^2 = (3 - 2x_1) + (3 - 2x_2) = 6 - 2 \times (x_1 + x_2) = 6 - 2 \times (-2) = \boxed{10}$$

This technique can be used repeatedly to continuously reduce the power and thus simplify a target expression. For example,

$$\begin{aligned}
x_1^4 + x_2^4 &= (x_1^2)^2 + (x_2^2)^2 \\
&= (3 - 2x_1)^2 + (3 - 2x_2)^2 \\
&= (9 - 12x_1 + 4x_1^2) + (9 - 12x_2 + 4x_2^2) \\
&= 18 - 12 \times (x_1 + x_2) + 4 \times (x_1^2 + x_2^2) \\
&= 18 - 12 \times (-2) + 4 \times 10 \qquad \text{by previous example} \\
&= \boxed{82}
\end{aligned}$$

Chapter 2: Vieta Theorem

What also deserves mentioning is that this technique can be used to tackle some asymmetric expressions which are usually difficult to be handled with other techniques. Some practice problems fall into this category.

2.3.3 Recursion

While the power reducing technique is able to handle higher power expressions in a relatively easier way than the polynomial transformation can, it may be still a bit challenging if the power is too high. It turns out that expressions in the form of $x_1^n + x_2^n$ can be calculated in a recursive fashion.

> **Theorem 2.3.1 Vieta's Theorem - Recursion**
>
> Suppose x_1 and x_2 are the two roots of quadratic equation $ax^2 + bx + c = 0$. Let $y_n = x_1^n + x_2^n$, where n is an integer. Then the following recursion always hold:
>
> $$ay_{n+2} + by_{n+1} + cy_n = 0 \qquad (2.2)$$
>
> where y_0 always equals 2 and y_1 equals $-\frac{b}{a}$ by Vieta's theorem[a].
>
> ---
> [a]This is because $y_0 = x_1^0 + x_2^0 = 1 + 1$ and $y_1 = x_1 + x_2$.

For example, the recursion relation in *Example 2.3.1* is

$$y_{n+2} + 2y_{n+1} - 3y_n = 0 \implies y_{n+2} = -2y_{n+1} + 3y_n$$

$$\begin{aligned}
\therefore y_0 &= x_1^0 + x_2^0 &&= 2 \\
y_1 &= x_1 + x_2 &&= -2 \\
y_2 &= x_1^2 + x_2^2 &&= -2 \times (-2) + 3 \times 2 &&= \boxed{10} \\
y_3 &= x_1^3 + x_2^3 &&= -2 \times 10 + 3 \times (-2) &&= -26 \\
y_4 &= x_1^4 + x_2^4 &&= -2 \times (-26) + 3 \times 10 &&= \boxed{82}
\end{aligned}$$

These answers agree with previous results.

In fact, the recursion relation (2.2) still holds even if n is a negative integer. As such, $\frac{1}{x_1} + \frac{1}{x_2}$ can also be computed as

$$y_{-1} = \frac{1}{3} \times (y_1 + 2y_0) = \frac{1}{3} \times (-2 + 2 \times 2) = \boxed{\frac{2}{3}}$$

This implies that expressions in the form of $\frac{1}{x_1^n} + \frac{1}{x_2^n}$ can be evaluated recursively too.

Proving *Theorem 2.3.1* is straightforward. Because x_1 and x_2 are roots of $ax^2 + bx + c = 0$, it must be true that

$$ax_1^2 + bx_1 + c = 0 \quad \text{and} \quad ax_2^2 + bx_2 + c = 0$$

Multiplying both sides of the above two relations with x_1^n and x_2^n, respectively, yields

$$ax_1^{n+2} + bx_1^{n+1} + cx_1^n = 0 \quad \text{and} \quad ax_2^{n+2} + bx_2^{n+1} + cx_2^n = 0$$

Adding these two equations and replacing $x_1^k + x_2^k$ with y_k where $k = n, n+1, n+2$ lead to the desired result immediately.

> Recursion (2.2) can be extended to equations having more than two variables.

For example, let $x_{1,2,3}$ be the three roots of $ax^3 + bx^2 + cx + d = 0$ and $y_n = x_1^n + x_2^n + x_3^n$, then

$$ay_{n+3} + by_{n+2} + cy_{n+1} + dy_n = 0$$

2.3.4 Equation Transformation

It may be possible to transform not only the target expression, but also the given equation. For example, if a new equation can be constructed with roots being $\frac{1}{x_1}$ and $\frac{1}{x_2}$, then the expression $\frac{1}{x_1} + \frac{1}{x_2}$ can be evaluated by applying Vieta's theorem on the new equation instead of the original one. Let's still use *Example 2.3.1* to illustrate this technique.

Because x_1 and x_2 are the two roots of equation
$$x^2 + 2x - 3 = 0,$$
$\frac{1}{x_1}$ and $\frac{1}{x_2}$ must be the roots of this equation
$$\left(\frac{1}{x}\right)^2 + 2 \times \left(\frac{1}{x}\right) - 3 = 0 \implies 3x^2 - 2x - 1 = 0$$

Therefore, by Vieta's theorem,
$$\frac{1}{x_1} + \frac{1}{x_2} = \boxed{\frac{2}{3}}$$

This agrees with previous result.

2.3.5 Equation Construction

Some problems can be solved by using the Vieta's theorem conversely.

> Given two numbers a and b, if both their sum s and product p are known, then they must be the two roots of this equation
> $$t^2 - st + p = 0$$

Let's consider the following example.

Example 2.3.2

Let x, y, and z be real numbers satisfying $x = 6 - y$ and $z^2 = xy - 9$. Show that $x = y$.

Solution

From the given conditions, we have
$$x + y = 6 \quad \text{and} \quad xy = z^2 + 9$$
Therefore x and y must be the two real roots of
$$u^2 - 6u + (z^2 + 9) = 0$$
It follows that this equation's discriminant must be non-negative:
$$\Delta = 36 - 4(z^2 + 9) \geq 0 \implies z^2 \leq 0 \implies z = 0 \implies \Delta = 0$$
This implies the equation has two equal roots, or $x = y$. When $x = y$, we find $(x, y, z) = (3, 3, 0)$ satisfies the given conditions. Therefore we conclude that $x = y$ must hold.

Done.

2.3.6 Properties of Roots

Vieta's theorem can also be used to determine some properties of the roots. The following problem is such an example.

Example 2.3.3

Find the range of real number m such that both roots of equation $x^2 - mx + m + 1 = 0$ are positive.

Solution

Let x_1 and x_2 be the two roots. If they both are positive, then
$$\begin{cases} x_1 + x_2 &= m & > 0 \\ x_1 \cdot x_2 &= m + 1 & > 0 \\ \Delta &= m^2 - 4 \times (m+1) & \geq 0 \end{cases} \implies m \geq 2 + 2\sqrt{2}$$

Done.

Chapter 2: Vieta Theorem

It is also possible to tackle the following cases:

i) One positive root and one negative root.

ii) Both roots are greater than 1.

iii) One root is greater than 1 and the other is less than 1.

iv) $\cdots$

Practice contains such problems.

> While Vieta's theorem holds even if roots are not real numbers, one must check discriminant if roots are known to be real.

2.4 Practice

Practice 1

Let a and b be the two roots of $x^2 + x + 1 = 0$. Evaluate $a^2 + b^2$. Try to find at least three different solutions without solving the equation directly.

Practice 2

Let a and b be the two roots of $x^2 + x + 1 = 0$. Evaluate
$$\frac{1}{a} + \frac{1}{b}$$
Try to find at least three different solutions without solving the equation directly.

Practice 3

If $m^2 = m + 1$, $n^2 - n = 1$ and $m \neq n$, evaluate:

$$(i) \quad m^5 + n^5 \quad \text{and} \quad (ii) \quad \frac{1}{m^5} + \frac{1}{n^5}$$

Practice 4

Let a and b be the two roots of $x^2 + x - 1 = 0$. Evaluate $a - b$.

Practice 5

If $x^2 + 11x + 16 = 0$, $y^2 + 11y + 16 = 0$, and $x \neq y$, what is the value of

$$\sqrt{\frac{x}{y}} - \sqrt{\frac{y}{x}}$$

Practice 6

Let x_1 and x_2 be two roots of $x^2 - x - 1 = 0$. Find the value of

$$2x_1^5 + 5x_2^3$$

Practice 7

If the difference of the two roots of the equation $x^2 + 6x + k = 0$ is 2, what is the value of k?

Chapter 2: Vieta Theorem

Practice 8

If one root of the equation $x^2 - 6x + m^2 - 2m + 5 = 0$ is 2. Find the value of the other root and m.

Practice 9

Three of the roots of $x^4 + ax^2 + bx + c = 0$ are 2, -3, and 5. Find the value of $a + b + c$.

Practice 10

Let x_1 and x_2 be the two roots of $x^2 - 3mx + 2(m-1) = 0$. If $\frac{1}{x_1} + \frac{1}{x_2} = \frac{3}{4}$, what is the value of m?

Practice 11

Let x_1 and x_2 be the two real roots of the equation

$$x^2 - 2(k+1)x + k^2 + 2 = 0$$

If $(x_1 + 1)(x_2 + 1) = 8$, find the value of k

Practice 12

Let α_n and β_n be two roots of equation $x^2 + (2n+1)x + n^2 = 0$ where n is a positive integer. Evaluate the following expression

$$\frac{1}{(\alpha_3 + 1)(\beta_3 + 1)} + \frac{1}{(\alpha_4 + 1)(\beta_4 + 1)} + \cdots + \frac{1}{(\alpha_{20} + 1)(\beta_{20} + 1)}$$

(Ref 1983 Shanghai)

Chapter 2: Vieta Theorem

Practice 13

If $a \neq 0$ and $\frac{1}{4}(b-c)^2 = (a-b)(c-a)$, compute $\frac{b+c}{a}$.
(Ref 1999 China)

Practice 14

Let real numbers a, b, c satisfy $a > 0$, $b > 0$, $2c > a+b$, and $c^2 > ab$. Prove

$$c - \sqrt{c^2 - ab} < a < c + \sqrt{c^2 - ab}$$

Practice 15

If both roots of $x^2 + ax + b + 1 = 0$ are positive integers, show that $a^2 + b^2$ cannot be a prime number.

Practice 16

Find integer m such that the equation $x^2 - mx + m + 1 = 0$ has two positive integer roots.

Practice 17

Find the range of real number a if the two roots of $x^2 + 2ax + 6 - a = 0$ satisfy each of the following conditions:

i) two roots are both greater than 1

ii) one root is greater than 1 and the other is less than 1

Chapter 2: Vieta Theorem

Practice 18

Suppose a_1, b_1, c_1, a_2, b_2, and c_2 are all positive real numbers. If both $a_1x^2 + b_1x + c_1 = 0$ and $a_2x^2 + b_2x + c_2 = $ are solvable in real numbers. Show that their roots must be all negative. Furthermore, prove equation $a_1a_2x^2 + b_1b_2x + c_1c_2 = 0$ has two negative real roots too.

Practice 19

Find the sum of all possible integer values of a such that the following equation is solvable in integers:
$$(a+1)x^2 - (a^2+1)x + (2a^2 - 6) = 0$$

Practice 20

In $\triangle ABC$, let a, b, and c be the lengths of sides opposite to $\angle A$, $\angle B$ and $\angle C$, respectively. D is a point on side AB satisfying $BC = DC$. If $AD = d$, show that
$$c + d = 2 \cdot b \cdot \cos A \quad \text{and} \quad c \cdot d = b^2 - a^2$$

Practice 21

If real numbers m and n satisfy $mn \neq 1$, $19m^2 + 99m + 1 = 0$ and $19 + 99n + n^2 = 0$, what is the value of
$$\frac{mn + 4m + 1}{n}$$

Chapter 2: Vieta Theorem

Practice 22

Let α and β be two real roots of $x^4 + k = 3x^2$ and also satisfy $\alpha + \beta = 2$. Find the value of k.

Practice 23

Let real numbers a, b, and c satisfy $a + b + c = 2$ and $abc = 4$. Find

i) the minimal value of the largest among a, b, and c.

ii) the minimal value of $|a| + |b| + |c|$.

(Ref 2003 China)

Practice 24

Compute the value of

$$\sqrt[3]{2 + \frac{10}{3\sqrt{3}}} + \sqrt[3]{2 - \frac{10}{3\sqrt{3}}}$$

and simplify

$$\sqrt[3]{2 + \frac{10}{3\sqrt{3}}} \quad \text{and} \quad \sqrt[3]{2 - \frac{10}{3\sqrt{3}}}$$

Chapter 2: Vieta Theorem

Practice 25

If all coefficients of the polynomial

$$f(x) = a_n x^n + a_{n-1} x^{n-1} + \cdots + a_3 x^3 + x^2 + x + 1$$

are real numbers, prove that its roots cannot be all real.

Chapter 3

High Degree Equation

3.1 Introduction

When $n \geq 3$, the close formed solution to the following equation usually does not exist or is too complex:

$$a_0 x^n + a_1 x^{n-1} + \cdots + a_{n-1} x + a_n = 0 \quad (a_0 \neq 0)$$

Having said this, high degree equation related problems appear in various competitions from time to time. Such equations are special cases and can be solved using elementary techniques. The key to tackle such problems is to identify features presented in a given equation and then to apply appropriate technique accordingly.

3.2 Factorization

Factorization is the method of choice to solve high degree equations. Effectively using this method relies on one's ability to factorize a given polynomial. This often requires experience and practice.

Chapter 3: High Degree Equation

Example 3.2.1

Solve this equation: $6x^3 - x^2 - 11x + 6 = 0$.

Solution

Because the sum of all coefficients equals zero, the given equation must be divisible by $(x-1)$. Then by polynomial division, we find

$$6x^3 - x^2 - 11x + 6 = (x-1)(6x^2 + 5x - 6)$$

Next, $6x^2 + 5x - 6$ can be factorized into $(3x-2)(2x+3)$, therefore

$$6x^3 - x^2 - 11x + 6 = (x-1)(3x-2)(2x+3) \implies x_{1,2,3} = \boxed{1, \frac{2}{3}, -\frac{3}{2}}$$

Done.

3.3 Rational Zero Theorem

To some extents, factorization is somewhat similar to guessing roots. This is because showing $(kx + m)$ divides

$$f(x) = a_0 x^n + a_1 x^{n-1} + \cdots + a_{n-1} x + a_n \quad (a_0 \neq 0)$$

is the same as stating $-\frac{m}{k}$ is one root of $f(x) = 0$.

When trying root guessing, it is easier to start by rational numbers[1]. If there exists a rational root, this root must satisfy the rational zero theorem which is stated below. Therefore this theorem can be used to list all the *possible* rational roots.

---

[1] Rational numbers include integers.

Chapter 3: High Degree Equation

> **Theorem 3.3.1 Rational Zero Theorem**
>
> Let $P(x) = a_0 x^n + a_1 x^{n-1} + \cdots + a_{n-1} x + a_n$ be a n-degree polynomial with integer coefficients. If $\frac{p}{q}$ is a root of $P(x) = 0$ where integer p and q are co-prime, then $p \mid a_n$ and $q \mid a_0$.

A corollary to this theorem states that

> If the first coefficient $a_0 = 1$, then all possible rational roots must be integers and they must divide the last coefficient a_n.

Let's look at one example.

Example 3.3.1

Solve this equation:

$$6x^3 + 13x^2 + x - 2 = 0$$

To employ the rational zero theorem, the first step is to list all the divisors of a_0 and a_n:

$$6 : \pm 1, \pm 2, \pm 3, \pm 6$$
$$-2 : \pm 1, \pm 2$$

Therefore, all the possible rational roots are

$$\pm 1, \pm \frac{1}{2}, \pm \frac{1}{3}, \pm \frac{1}{6}, \pm 2, \pm \frac{2}{3}$$

Setting these values to the original equation finds the following three roots:

$$\boxed{-\frac{1}{2}, \frac{1}{3}, -2}$$

This result also means that the original equation can be factorized into

$$6x^3 + 13x^2 + x - 2 = (2x+1)(3x-1)(x+2)$$

Chapter 3: High Degree Equation

3.4 Substitution

Substitution is another widely used technique to solve high degree equations. The objective of substitution is to simplify a given equation to one that can be easily solved. For example, this 4th degree equation
$$ax^4 + bx^2 + c = 0$$
can be transformed to a quadratic one by substituting $y = x^2$:
$$ay^2 + by + c = 0 \qquad (3.1)$$
This latter form is clearly solvable. Upon having solved y, solutions to the original equation can be easily computed as well.

Substitution can go further. Let's consider the following example. It is difficult to solve by factorization because its roots are irrational and thus is almost impossible to guess. However, it can be transformed to the form of *(3.1)* by a simple linear transformation.

Example 3.4.1

Solve this equation
$$x^4 + 2x^3 - 3x^2 - 4x + 3 = 0$$

Solution

Let $x = y + k$ where k is a to-be-determined constant. Hence
$$(y+k)^4 + 2(y+k)^3 - 3(y+k)^2 - 4(y+k) + 3 = 0$$
In order to transform it to the form of *(3.1)*, let's set its coefficient of y^3 to zero:
$$4k + 2 = 0 \implies k = -\frac{1}{2}$$
Substitute $k = -\frac{1}{2}$ in y's coefficient:
$$4k^3 + 6k^2 - 6k - 4 = 4 \times \left(-\frac{1}{2}\right)^3 + 6 \times \left(-\frac{1}{2}\right)^2 - 6 \times \left(-\frac{1}{2}\right) - 4 = 0$$

Therefore, by substituting $x = y - \frac{1}{2}$, both y^3 and y are canceled:

$$y^4 - \frac{9}{2}y^2 + \frac{65}{16} = 0 \Leftrightarrow 16y^4 - 72y^2 + 65 = 0$$

$$\therefore \ y^2 = \frac{5}{4}, \frac{13}{4} \implies x_{1,2,3,4} = \boxed{\frac{\pm\sqrt{5} - 1}{2}, \frac{\pm\sqrt{13} - 1}{2}}$$

Done.

There exist various types of substitutions. Linear substitution such as the one used in the preceding example is just one of them. Finding an appropriate substitution is the core of this technique. It requires careful observation of features which are exhibited by the given equations. Additional techniques will be illustrated in some practice problems.

3.5 $n + 1$ Roots

The fundamental theorem of algebra states that

> **Theorem 3.5.1 Fundamental Theorem of Algebra**
>
> Every non-zero, single variable, n-degree polynomial with complex coefficients has exactly n complex roots.

For example, an quadratic equation always have two complex roots if equal roots are counted as two. There are several corollaries can be drawn from this theorem. Two of the most used ones are

i) If an n−degree equation has $n + 1$ roots, then this equation must always equal zero (i.e. all coefficients are zeros).

ii) If two n−degree equations have the same values at $n + 1$ distinct points, then they must be equivalent.

Chapter 3: High Degree Equation

To clarify, the 2nd corollary above means that for two n−degree polynomials $f(x)$ and $g(x)$, if there exist $n+1$ distinct x_i, ($i = 1, 2, \cdots, n, n+1$), such that $f(x_i) = g(x_i)$, then these two polynomials are equivalent.

This conclusion holds because if $f(x)$ is subtracted from $g(x)$, the result equation is at most n−degree. However, it will has $n+1$ roots which means this difference equation must always equal zero. So the two original ones must be equivalent.

Example 3.5.1

Let polynomials $P(x)$, $Q(x)$, $R(x)$, and $S(x)$ satisfy:

$$P(x^5) + xQ(x^5) + x^2 R(x^5) = (x^4 + x^3 + x^2 + x + 1)S(x)$$

Show that $P(1) = Q(1) = R(1) = 0$.

This problem appears to be quite abstract. Let's use the fundamental theory of algebra to tackle it[2].

Proof

Let $\omega_1, \omega_2, \omega_3$, and ω_4 are the four zeros of $x^4 + x^3 + x^2 + x + 1$.

$\therefore \quad x^4 + x^3 + x^2 + x + 1 = (x - \omega_1)(x - \omega_2)(x - \omega_3)(x - \omega_4)$

It is clear that $\omega_i^5 = 1$, for $i = 1, 2, 3, 4$. Therefore by the given condition, setting $x = \omega_1, \omega_2$, and ω_3, respectively, yields

$$P(1) + \omega_1 Q(1) + \omega_1^2 R(1) = 0$$

$$P(1) + \omega_2 Q(1) + \omega_2^2 R(1) = 0$$

$$P(1) + \omega_3 Q(1) + \omega_3^2 R(1) = 0$$

---

[2]Though th problem does not explicitly state coefficients of these polynomials are complex numbers, we can safely assume so in pre-college level competitions.

This implies that ω_1, ω_2, and ω_3 are three distinct roots to the quadratic equation

$$R(1) \cdot x^2 + Q(1) \cdot x + P(1) = 0$$

which is impossible unless $R(1) = Q(1) = P(1) = 0$.

$$QED$$

3.6 Practice

Practice 1

Solve $x^3 - 3x + 2 = 0$.

Practice 2

Solve $x^4 - 2x^3 - 7x^2 + 8x + 12 = 0$.

Practice 3

Find the range of real number a if equation $\left| \frac{x^2}{x-1} \right| = a$ has exactly two distinct real roots.

(Ref 2006 China)

Practice 4

Solve this equation: $(x^2 - x - 1)^{x+2} = 1$.

Chapter 3: High Degree Equation

Practice 5

Solve this system:
$$\begin{cases} \dfrac{4}{3x-2y} + \dfrac{3}{2x-5y} = 10 \\[2mm] \dfrac{5}{3x-2y} - \dfrac{2}{2x-5y} = 1 \end{cases}$$

Practice 6

Solve this equation $2x^4 + 3x^3 - 16x^2 + 3x + 2 = 0$.

Practice 7

Solve this equation $(x-2)(x+1)(x+4)(x+7) = 19$.

Practice 8

Solve equation $(6x+7)^2(3x+4)(x+1) = 6$ in real numbers.

Practice 9

If all roots of the equation

$$x^4 - 16x^3 + (81-2a)x^2 + (16a-142)x + (a^2 - 21a + 68) = 0$$

are integers, find the value of a and solve this equation.
(Ref 2009 Jiang Xi)

Chapter 3: High Degree Equation

Practice 10

Suppose the graph of $f(x) = x^4 + ax^3 + bx^2 + cd + d$, where a, b, c, d are all real constants, passes through three points $A(2, \frac{1}{2})$, $B(3, \frac{1}{3})$, and $C(4, \frac{1}{4})$. Find the value of $f(1) + f(5)$.

Practice 11

Find a quadratic polynomial $f(x) = x^2 + mx + n$ such that

$$f(a) = bc, \quad f(b) = ca, \quad f(c) = ab$$

where a, b, c are three distinct real numbers.

Practice 12

Let $f(x) = 2016x - 2015$. Solve this equation

$$\underbrace{f(f(f(\cdots f(x))))}_{\text{2017 iterations}} = f(x)$$

Chapter 3: High Degree Equation

Chapter 4

Non-Polynomial Equation

Non-polynomial equations can appear in many different forms. It is hard to enumerate all the possibilities. However, the vast majority of such equations can be de-composed to several basic types. This chapter will cover those most commonly seen ones and their corresponding solving techniques. These basic techniques may appear to be simple, but can be combined to solve certain challenging competition problems.

4.1 Absolute Value

Solving an equation involving absolute values often requires casework. The goal is to eliminate absolute value operators and thus transform the given equation to a regular polynomial one.

Let's consider the following example:

Example 4.1.1

Solve this equation $\mid x-1 \mid + \mid x-2 \mid = 3$.

Chapter 4: Non-Polynomial Equation

Solution

The first step is to find the zero points of all the terms involving absolute operators. In this example, there are two:

$$|x-1|=0 \implies x=1$$

$$|x-2|=0 \implies x=2$$

This means totally 3 different cases need to be examined:

```
      case 1        case 2        case 3
─────────●─────────────●──────────────→
         x = 1         x = 2
```

i) if $x \leq 1$, the original equation becomes

$$-(x-1)-(x-2)=3 \implies x=0$$

ii) if $1 < x \leq 2$, the original equation becomes

$$(x-1)-(x-2)=3 \implies \text{no solution}$$

iii) if $2 < x$, the original equation becomes

$$(x-1)+(x-2)=3 \implies x=3$$

Hence, we find the given equation has two solutions $\boxed{0, 3}$.

Done.

Sometimes, plotting a graph can help visualize and understand a function's overall properties. Usually, the graph of $|f(x)|$ can be obtained from that of $f(x)$ by flipping the part below x-axis over to above.

Chapter 4: Non-Polynomial Equation

Example 4.1.2

Plot the graph of function $f(x) = |x-1| + |x-2|$.

Solution

Because $f(x)$ can be viewed as a sum of two sub-functions, it will be more convenient to plot these two sub-functions first and then add them together.

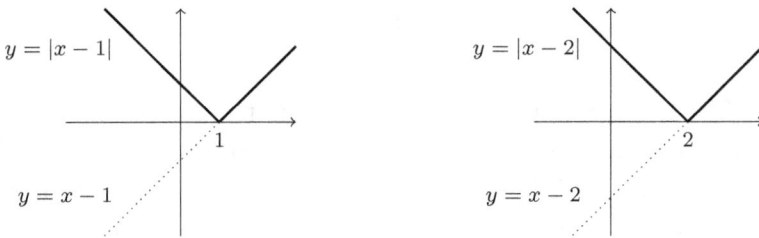

Adding these two together gives us the graph of

$$f(x) = |x-1| + |x-2|$$

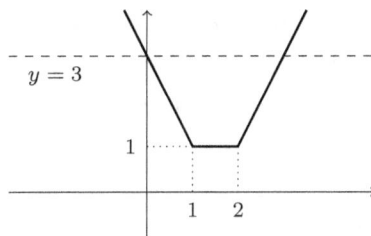

This graph also confirms that $f(x) = 3$ will have two solutions which agrees with the conclusions of *Example 4.1.1*.

Done.

Chapter 4: Non-Polynomial Equation

4.2 Irrational

Irrational equations often appear in the following form:
$$\sqrt{\cdots} + \sqrt{\cdots} + \cdots = \cdots$$

Typical techniques to solve such equations include:

i) the square method

ii) the simplification method

The square method is to re-arrange the terms so that at least one radical operator can be eliminated by taking squares on both sides of the equation. This method works well when the number of radical terms is small. If there is only one radical operator, taking square once will eliminate it. However if there are more radical terms, several iterations of squaring may be required. Consequently, the original equation may turn to a high degree equation which itself may be difficult to solve.

Radical terms may be nested. It is always beneficial to first examine whether nested radicals can be simplified. Simplifying nested radical is discussed in the book *Power Calculation*. Here is an example.

Example 4.2.1

Solve this equation
$$\sqrt{x+3-4\sqrt{x-1}} + \sqrt{x+8-6\sqrt{x-1}} = 1$$

Solution

Notice that both nested radical terms can be simplified. Hence, this equation can be simplified to:
$$\mid \sqrt{x-1} - 2 \mid + \mid \sqrt{x-1} - 3 \mid = 1$$

This leads to an equation involving two absolute value terms.

i) When $\sqrt{x-1} < 2$, then:
$$-(\sqrt{x-1}-2) - (\sqrt{x-1}-3) = 1 \implies \sqrt{x-1} = 2$$
This contradicts the assumption that $\sqrt{x-1} < 2$.

ii) When $2 \leq \sqrt{x-1} < 3$, then:
$$(\sqrt{x-1}-2) - (\sqrt{x-1}-3) = 1$$
This is an identity which implies any x satisfying the assumption will be a solution. Or
$$2 \leq \sqrt{x-1} < 3 \implies 5 \leq x < 10$$

iii) When $3 \leq \sqrt{x-1}$, then:
$$(\sqrt{x-1}-2) + (\sqrt{x-1}-3) = 1 \implies x = 10$$

Therefore we conclude that the solution is $\boxed{5 \leq x \leq 10}$.

Done.

4.3 Fraction

Equations involving fraction terms frequently appear as a system like the following two examples.

Example 4.3.1

Let real numbers $x, y,$ and z satisfy
$$x + \frac{1}{y} = 4 \quad, \quad y + \frac{1}{z} = 1 \quad, \quad z + \frac{1}{x} = \frac{7}{3}$$
Find the value of xyz.

Chapter 4: Non-Polynomial Equation

Solution

This problem, though involves fraction terms, can be solved using the regular cancellation technique.

Let's construct an equation of x by canceling y and z.

$$4 = x + \frac{1}{y} = x + \frac{1}{1-\frac{1}{z}} = x + \frac{1}{1-\frac{1}{\frac{7}{3}-\frac{1}{x}}}$$

This equation can be simplified to

$$4x^2 - 12x + 9 = 0 \Leftrightarrow (2x-3)^2 = 0 \implies x = \frac{3}{2}$$

It follows that $z = \frac{7}{3} - \frac{1}{x} = \frac{5}{3}$ and $y = 1 - \frac{1}{z} = \frac{2}{5}$.

$$\therefore \quad xyz = \frac{3}{2} \times \frac{2}{5} \times \frac{5}{3} = \boxed{1}$$

Done.

When the conditions are given in the form of

$$\frac{x_1}{y_1} = \frac{x_2}{y_2} = \frac{x_3}{y_3} = \cdots,$$

a useful technique is to set the ratio to k and then transform the conditions to be relations with respect to k. Sometimes, the following identity may help simplify the calculation:

$$\frac{x_1}{y_1} = \frac{x_2}{y_2} \implies \frac{x_1}{y_1} = \frac{x_2}{y_2} = \frac{x_1 \pm x_2}{y_1 \pm y_2} \qquad (4.1)$$

Let's look at this example.

Example 4.3.2

Let non-zero real numbers a, b, c satisfy $a + b + c \neq 0$. If the following relations hold

$$\frac{a+b-c}{c} = \frac{a-b+c}{b} = \frac{-a+b+c}{a}$$

Find the value of

$$\frac{(a+b)(b+c)(c+a)}{abc}$$

(Ref 2004 Tian Jin)

Solution

Suppose

$$\frac{a+b-c}{c} = \frac{a-b+c}{b} = \frac{-a+b+c}{a} = k$$

Then

$$k = \frac{(a+b-c) + (a-b+c) + (-a+b+c)}{a+b+c} = 1$$

Therefore

$$\frac{a+b}{c} = \frac{a+c}{b} = \frac{b+c}{a} = k+1 = 2$$

Finally,

$$\frac{(a+b)(b+c)(c+a)}{abc} = 2 \times 2 \times 2 = \boxed{8}$$

Done.

Chapter 4: Non-Polynomial Equation

4.4 Floor and Ceiling

Equations involving floor or ceiling often appear challenging. Careful observation and analysis are keys to solve such problems. The focus is how to remove floor and ceiling operators.

Let's consider the following example where floor function $\lfloor x \rfloor$ represents the largest integer not exceeding real number x.

Example 4.4.1

How many solutions does the following system have?

$$\begin{cases} \lfloor x \rfloor + 2y = 1 \\ \lfloor y \rfloor + x = 2 \end{cases}$$

(Ref 2015 Hua Bei)

Solution

Because $\lfloor y \rfloor$ represents an integer, the 2nd equation implies x is an integer which means $\lfloor x \rfloor = x$. Then, by the 1st equation, y is either an integer or its decimal part is 0.5.

If y is an integer, then $\lfloor y \rfloor = y$. Hence

$$\begin{cases} x + 2y = 1 \\ y + x = 2 \end{cases} \Longrightarrow \begin{cases} x = 3 \\ y = -1 \end{cases}$$

If $y = y_0 + 0.5$ where y_0 is an integer, then

$$\begin{cases} x + 2 \times (y_0 + 0.5) = 1 \\ y_0 + x = 2 \end{cases} \Longrightarrow \begin{cases} x = 4 \\ y_0 = -2 \end{cases} \Longrightarrow y = -1.5$$

In conclusion, this system has $\boxed{2}$ solutions.

Done.

Chapter 4: Non-Polynomial Equation

Similar to solving some other types of problems, plotting function graphs can be helpful in solving floor and ceiling related equations. For instance, the essential step to solve the 14th problem in 2015 AIME I is to plot the function $y = x\lfloor\sqrt{x}\rfloor$.

Geometrically, floor and ceiling related functions are usually piecewise. This means that they may not be smooth and, sometimes, not even be continuous. Let's start with the simplest: plotting $\lfloor x \rfloor$ and $\lceil x \rceil$.

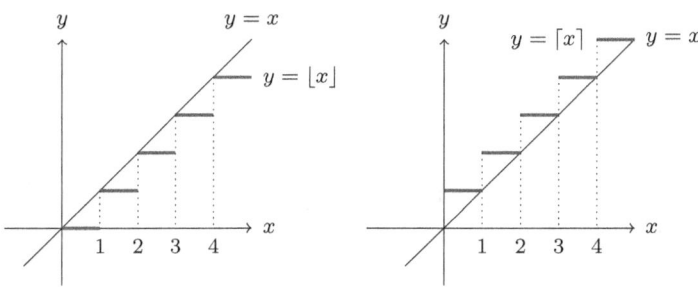

Both graphs are easy to understand. This is because their value stay the same until x reaches the next whole integer at which point y will make a jump. Therefore the plot exhibits a step-like shape.

Example 4.4.2

Plot the function $y = x\lfloor x \rfloor$ when $x \geq 0$.

Solution

While $y = x$ is continuous, $y = \lfloor x \rfloor$ is step-like. As a result, function $y = x\lfloor x \rfloor$ is likely not continuous and thus requires piecewise analysis.

- When $0 \leq x < 1$, $\lfloor x \rfloor = 0 \implies y = 0$.
- When $1 \leq x < 2$, $\lfloor x \rfloor = 1 \implies y = x$.
- When $2 \leq x < 3$, $\lfloor x \rfloor = 2 \implies y = 2x$.

Chapter 4: Non-Polynomial Equation

- When $3 \leq x < 4$, $\lfloor x \rfloor = 3 \implies y = 3x$.

Therefore, the graph of this function is made up of many unit-length segments: $[0, 1), [1, 2), \ldots$. Within each segment, the graph is a straight line, with their slopes increasing gradually.

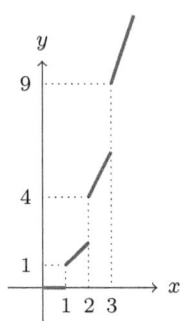

Done.

Plotting function $y = x\lfloor \sqrt{x} \rfloor$ is similar to that of *Example 4.4.2* except the segments will be $[0, 1^2), [1^2, 2^2), [2^2, 3^2), \ldots$, and so on.

4.5 Practice

Practice 1

Let positive numbers a, b, c, d, e, f satisfy $\frac{bcdef}{a} = 4$, $\frac{acdef}{b} = 9$, $\frac{abdef}{c} = 16$, $\frac{abcef}{d} = \frac{1}{4}$, $\frac{abcdf}{e} = \frac{1}{9}$, and $\frac{abcde}{f} = \frac{1}{16}$. Compute the value of $(a + c + e) - (b + d + f)$.

Chapter 4: Non-Polynomial Equation

Practice 2

If $\frac{ab}{a+b} = \frac{1}{15}$, $\frac{bc}{b+c} = \frac{1}{17}$, $\frac{ca}{c+a} = \frac{1}{16}$, find the value of

$$\frac{abc}{ab + bc + ca}$$

(Ref Tai Yuan)

Practice 3

Solve this system:

$$\begin{cases} x_1 + x_2 = x_2 + x_3 = \cdots = x_{1997} + x_{1998} = x_{1998} + x_{1999} = 1 \\ x_1 + x_2 + x_3 + \cdots + x_{1998} + x_{1999} = 1999 \end{cases}$$

(Ref 1999 Hua Bei)

Practice 4

Let a, b, and c be three distinct numbers such that

$$\frac{a+b}{a-b} = \frac{b+c}{2(b-c)} = \frac{c+a}{3(c-a)}$$

Prove that $8a + 9b + 5c = 0$.

Practice 5

Solve this system:

$$\begin{cases} |x+y| + |x| = 4 \\ 2|x+y| + 3|x| = 9 \end{cases}$$

Chapter 4: Non-Polynomial Equation

Practice 6

Solve this system:
$$\begin{cases} |x+y| = 1 \\ |x| + 2|y| = 3 \end{cases}$$

Practice 7

Let x, y, z be three integers satisfying
$$\begin{cases} |x+y| + |y+z| + |z+x| = 4 \\ |x-y| + |y-z| + |z-x| = 2 \end{cases}$$

Compute $x^2 + y^2 + z^2$.

Practice 8

Let x be a positive number. Denote by $\lfloor x \rfloor$ the integer part of x and by $\{x\}$ the decimal part of x. Find the sum of all positive numbers satisfying $5\{x\} + 0.2\lfloor x \rfloor = 25$.

Practice 9

If $abc = 1$, solve this equation
$$\frac{2ax}{ab+a+1} + \frac{2bx}{bc+b+1} + \frac{2cx}{ca+c+1} = 1$$

Practice 10

Solve this system
$$\begin{cases} |x| + y = 12 \\ x + |y| = 6 \end{cases}$$

(Ref 2007 China)

Practice 11

Solve this equation in real numbers:

$$\sqrt{x} + \sqrt{y-1} + \sqrt{z-2} = \frac{1}{2} \times (x + y + z)$$

Practice 12

Let a, b, and c be the lengths of $\triangle ABC$'s three sides. Compute the area of $\triangle ABC$ if the following relations hold:

$$\frac{2a^2}{1+a^2} = b, \qquad \frac{2b^2}{1+b^2} = c, \qquad \frac{2c^2}{1+c^2} = a$$

Practice 13

Find $\lfloor x \rfloor$ where $x = 1 + \frac{1}{\sqrt{2}} + \frac{1}{\sqrt{3}} + \cdots + \frac{1}{\sqrt{10000}}$

Chapter 4: Non-Polynomial Equation

Chapter 5

Sequence

5.1 Introduction

A sequence is a series of ordered elements. The number of elements can be either limited or unlimited. The first term is usually denoted as a_1, though sometime a_0 is also used.

There are two basic questions when working on sequence related problems:

1. How to compute its nth term, a_n?

2. How to compute the sum of its first n terms, S_n?

The fundamental relation between a_n and S_n is

$$a_n = S_n - S_{n-1} \quad (n > 1) \tag{5.1}$$

The book *Power Calculation* by the same author has discussed some results and techniques in this subject. This chapter is to provide a more complete coverage of this subject with a special focus on more challenging problems and their corresponding solutions.

5.2 Basic Sequence

Arithmetic sequence and geometric sequence are two basic types. They are taught in classrooms and lay down the foundation for studying other more complex types of sequences.

5.2.1 Arithmetic Sequence

A sequence $\{a_n\}$ is said to be arithmetic if it satisfies

$$a_{n+1} = a_n + d \tag{5.2}$$

where d is a constant and is referred as common difference.

Given the first term a_1 and common difference d of an arithmetic sequence, its nth term can be written as

$$a_n = a_1 + (n-1)d \tag{5.3}$$

This relation can be generalized to the following one which holds for two arbitrary terms a_i and a_j:

$$a_j = a_i + (j-i)d \tag{5.4}$$

Sum of the first n terms of an arithmetic sequence is given by

$$S_n = \frac{(a_1 + a_n)n}{2} \tag{5.5}$$

In another word, S_n always equals half of the sum of the first and the last term multiplying the number of terms. This holds true for any consecutive terms which may not necessarily start from the beginning of the sequence.

$$S_{i,j} = a_i + a_{i+1} + a_{i+2} + \cdots + a_j = \frac{(a_i + a_j) \times (j - i + 1)}{2} \tag{5.6}$$

Chapter 5: Sequence

Example 5.2.1

Show that the sum of first n odd integers equals n^2.

Solution

The first n odd numbers form an arithmetic sequence whose initial term is 1 and common difference is 2. Therefore, by *(5.3)*, the last term equals

$$a_n = 1 + (n-1) \times 2 = 2n - 1$$

Then by *(5.5)*,

$$S_n = \frac{(1 + (2n-1)) \times n}{2} = n^2$$

Done.

For any arithmetic sequence, it always hold that

> For any odd number of consecutive terms in a given arithmetic sequence, the middle one equals the average of these terms.

This is followed by *(5.7)* which states the sum of any odd number of terms equals the product of its middle term and the number of terms.

$$a_1 + a_2 + \cdots + a_{2n-1} = a_n \times n \tag{5.7}$$

A relatively less known but useful conclusion to solve some competition problems is that sum of any n consecutive terms must be in a quadratic form of n. This is described below and can be proved by substituting a_n in *(5.5)* using *(5.3)*.

Chapter 5: Sequence

> **Theorem 5.2.1 Sum of Arithmetic Sequence**
>
> The sum of any n consecutive terms in an arithmetic sequence must be in the following form where A and B are two constants:
> $$S_n = An^2 + Bn$$

Let's consider the following example:

Example 5.2.2

Let the sum of first n terms of arithmetic sequence $\{a_n\}$ be S_n, and the sum of first n terms of another arithmetic sequence $\{b_n\}$ be T_n. If $S_n : T_n = 2n : (3n+7)$, compute the value of $a_8 : b_6$.
(Ref 2013 China)

Solution

Let $S_n = An^2 + Bn$ and $T_n = Cn^2 + Dn$ where A, B, C and D are all constants. Then

$$\frac{S_n}{T_n} = \frac{An^2 + Bn}{Cn^2 + Dn} = \frac{An + B}{Cn + D}$$

By the given condition $S_n : T_n = 2n : (3n+7)$, we can conclude

$$A = 2k, B = 0, C = 3k, D = 7k$$

where k is a constant.

It follows that

$$\frac{a_8}{b_6} = \frac{S_8 - S_7}{T_6 - T_5} = \frac{(2k \times 8^2) - (2k \times 7^2)}{(3k \times 6^2 + 7k \times 6) - (3k \times 5^2 + 7k \times 5)} = \boxed{\frac{3}{4}}$$

Done.

5.2.2 Geometric Sequence

A sequence $\{a_n\}$ is said to be geometric if it satisfies:

$$a_{n+1} = a_n \cdot q \qquad (5.8)$$

where q is a constant and is referred as common ratio.

Given the 1st term and common ratio, the nth term of a geometric sequence is given by

$$a_n = a_1 \cdot q^{n-1} \qquad (5.9)$$

Correspondingly, the relation between any two terms a_i and a_j can be written as:

$$a_j = a_i \cdot q^{j-i} \qquad (5.10)$$

The sum of the first n terms equals:

$$S_n = a_1 \times \frac{1-q^n}{1-q} \quad (q \neq 1) \qquad (5.11)$$

Clearly, when $q = 1$, this sequence becomes a constant sequence[1]. The sum of any n terms can be obtained by multiplying this constant value by count n.

Similarly, the sum of consecutive terms starting from a_i and ending at a_j can be computed as

$$S_{i,j} = a_i + a_{i+1} + \cdots + a_j = a_i \times \frac{1-q^{j-i+1}}{1-q} \qquad (5.12)$$

While the sum of an infinite number of terms in an arithmetic sequence is usually infinite, the sum of an infinite number of geometric sequence terms may exist if certain conditions are met. By

---

[1] A constant sequence is also an arithmetic sequence. The sum of arithmetic sequence still works.

Chapter 5: Sequence

(5.11), if $|q| < 1$, the term q^n will diminish when n becomes infinitely large. This leads to the following result:

$$S_\infty = a_1 + a_2 + \cdots = \frac{a_1}{1-q} \qquad (|q| < 1) \qquad (5.13)$$

It follows that if $|x| < 1$, we have

$$\frac{1}{1-x} = 1 + x + x^2 + x^3 + \cdots \qquad (|x| < 1) \qquad (5.14)$$

For advanced readers who have calculus knowledge, differentiating both sides of this relation k times yields

$$\frac{1}{(1-x)^k} = 1 + C_k^{k-1} x + C_{k+1}^{k-1} x^2 + C_{k+2}^{k-1} x^3 + \cdots \qquad (|x| < 1) \quad (5.15)$$

5.3 Linear Recursion

Sometimes, it is more convenient to define a sequence in a recursive fashion. Problems related to such types of sequences appear frequently in various math competitions.

In presence of a recursively defined sequence, the term of *solving the sequence* usually means to find an explicit formula for a_n. Such formula should only depend on n, and is unrelated to any previous terms.

5.3.1 $a_{n+1} = pa_n + q$

This is the simplest form of linear recursion that may appear in competitions. Its definition has the following characteristics:

i) Only two terms present in the recursion.

ii) Both p and q are constants.

Please note, while the two terms are usually consecutive, the solution presented in this section can tackle cases when they are not consecutive. When the two terms are consecutive, only one initial condition is needed. Otherwise more initial values will be required.

For example, given $a_{n+1} = 2a_n + 1$ and $a_1 = 1$, the whole sequence is completely defined. However, if the recursion is $a_{n+2} = 2a_n + 1$ and only one initial condition $a_1 = 1$ is given, this sequence is not completely defined because there is no way to determine a_2, a_4, and so on. We will need another initial condition such as $a_2 = 2$ in order to completely define this sequence.

Regardless of whether given terms are consecutive or not, this type of recursion can always be solved in a similar manner.

Let's consider the following example.

Example 5.3.1

Solve $a_{n+1} = 2a_n + 1$ where $a_1 = 1$.

Such recursion can always be transformed to a geometric sequence by substitution. Let's see how to achieve this.

Solution

The goal is to transform the given recursion to the following form:
$$(a_{n+1} + \alpha) = \beta(a_n + \alpha)$$
where α and β are two to-be-determined constants.

Expanding the above expression and rearranging the terms yield
$$a_{n+1} = \beta a_n + \alpha(\beta - 1)$$

Comparing the corresponding coefficients between this expanded form and the original recursion implies:

Chapter 5: Sequence

$$\begin{cases} \beta = 2 \\ \alpha(\beta - 1) = 1 \end{cases} \implies \begin{cases} \alpha = 1 \\ \beta = 2 \end{cases}$$

Therefore the given recursion is equivalent to

$$(a_{n+1} + 1) = 2 \times (a_n + 1)$$

Now let $b_n = a_n + 1$. It follows that $b_{n+1} = 2b_n$ which means $\{b_n\}$ is a geometric sequence. Its common ratio is 2 and its initial term is $b_1 = a_1 + 1 = 2$. Hence we have

$$b_n = 2 \times 2^{n-1} = 2^n \implies a_n = b_n - 1 = \boxed{2^n - 1} \qquad (5.16)$$

Done.

It is easy to verify that the result given in (5.16) does give each term correctly even though it is no longer recursively defined.

n	$a_{n+1} = 2a_n + 1$	$2^n - 1$
1	$a_1 = 1$ (initial value)	$2^1 - 1 = 1$
2	$a_2 = 2a_1 + 1 = 3$	$2^2 - 1 = 3$
3	$a_3 = 2a_2 + 1 = 7$	$2^3 - 1 = 7$
...		

> A recursion in the form of $a_{n+1} = pa_n + q$, where p and q are constants, can be solved by transforming it to a geometric sequence using substitution.

5.3.2 $a_{n+2} + pa_{n+1} + qa_n = 0$

This is a well-studied type of linear recursion which is often referred as being *homogeneous*. It usually involves three consecutive terms even though more terms may be present[2]. However, providing its recursion meets the following criteria, it can be solved using the technique discussed in this section.

i) All present terms, a_k, must be linear (i.e. first power)

ii) All coefficients must be constants

iii) There is no constant term (i.e. not in the form of $a_{n+2} + pa_{n+1} + qa_n + k = 0$ where $k \neq 0$)

One of the most famous sequences defined as a homogeneous linear sequence is the *Fibonacci* sequence

$$F_{n+2} = F_{n+1} + F_n \quad \text{and} \quad F_1 = F_2 = 1$$

By this recursive definition, its first few terms can be computed as $1, 1, 2, 3, 5, 8, 13, \cdots$. Meanwhile, it is possible to solve such a sequence so that F_n can be computed directly.

> **Theorem 5.3.1 Solve Homogeneous Linear Recursion**
>
> Suppose the given recursion is
>
> $$a_{n+2} + pa_{n+1} + qa_n = 0 \tag{5.17}$$
>
> and two initial values a_1 and a_2. Then it must hold that
>
> $$a_n = C_1 x_1^n + C_2 x_2^n \quad \text{if } x_1 \neq x_2$$
>
> or
>
> $$a_n = (C_1 + C_2 n) x_1^n \quad \text{if } x_1 = x_2$$
>
> where $C_{1,2}$ are two constants, and $x_{1,2}$ are roots of equation
>
> $$x^2 + px + q = 0 \tag{5.18}$$

---

[2]Non-consecutive terms can be viewed as the coefficients of those missing terms are zero. For example, $a_{n+3} + a_{n+2} + a_n = 0$ is equivalent to $a_{n+3} + a_{n+2} + 0 \times a_{n+1} + a_n = 0$. The latter involves four consecutive terms.

Chapter 5: Sequence

(5.18) is called the *characteristic equation* of (5.17). It can be obtained by replacing term a_{n+k} with x^k. The two constant coefficients C_1 and C_2 can be solved by plugging x_1 and x_2 with the two initial conditions. This is illustrated in the following example.

Example 5.3.2

Solve Fibonacci sequence which is defined as $F_{n+2} = F_{n+1} + F_n$, $F_1 = F_2 = 1$.

Step 1: Obtain the characteristic equation. In this example, it is

$$x^2 = x + 1$$

Step 2: Solve this characteristic equation.

$$x_1 = \frac{1+\sqrt{5}}{2}, x_2 = \frac{1-\sqrt{5}}{2}$$

Step 3: Therefore the solution is

$$F_n = C_1 \times \left(\frac{1+\sqrt{5}}{2}\right)^n + C_2 \times \left(\frac{1-\sqrt{5}}{2}\right)^n$$

where C_1 and C_2 are to-be-determined constants.

Step 4: Solve C_1 and C_2 using the two initial conditions.

$$\begin{cases} F_1 = C_1 \times (\frac{1+\sqrt{5}}{2})^1 + C_2 \times (\frac{1-\sqrt{5}}{2})^1 = 1 \\ F_2 = C_1 \times (\frac{1+\sqrt{5}}{2})^2 + C_2 \times (\frac{1-\sqrt{5}}{2})^2 = 1 \end{cases}$$

$$\Rightarrow C_1 = \frac{1}{\sqrt{5}} \text{ and } C_2 = -\frac{1}{\sqrt{5}}$$

Hence, the answer is

$$\boxed{F_n = \frac{1}{\sqrt{5}} \times \left(\frac{1+\sqrt{5}}{2}\right)^n - \frac{1}{\sqrt{5}} \times \left(\frac{1-\sqrt{5}}{2}\right)^n}$$

Chapter 5: Sequence

While the solution has irrational numbers, it can be verified that it indeed gives correct n^{th} term F_n.

The same solution can be used to solve recursions which involve more than three terms. For example, given
$$a_{n+3} + pa_{n+2} + qa_{n+1} + ra_n = 0 \tag{5.19}$$
its solution will be
$$a_n = C_1 x_1^n + C_2 x_2^n + C_3 x_3^n$$
where $C_{1,2,3}$ are three constants, and $x_{1,2,3}$ are roots of
$$x^3 + px^2 + qx + r = 0$$
Because *(5.19)* implies a term in sequence $\{a_n\}$ depends on three previous terms, three initial conditions $a_{1,2,3}$ must be given. They can be used to determine the three constants $C_{1,2,3}$.

The relation between a sequence's recurrence definition and its corresponding characteristic equation can also be used conversely. Let's consider the following example.

Example 5.3.3

Let real numbers a, b, x, y satisfy
$$\begin{cases} ax + by &= 3 \\ ax^2 + by^2 &= 7 \\ ax^3 + by^3 &= 16 \\ ax^4 + by^4 &= 42 \end{cases}$$

Find $ax^5 + by^5$.

Solution

Let $x + y = m$ and $xy = n$. Then x and y are the two roots of $t^2 - mt + n = 0$. This is the characteristic equation of sequence $\{S_n\}$ which is defined as following:
$$S_{n+2} - mS_{n+1} + nS_n = 0, \quad (n \geq 3)$$

Its general solution is
$$S_n = ax^n + by^n$$
and initial values are $S_1 = ax + by = 3$ and $S_2 = ax^2 + by^2 = 7$.

By recursion, we have
$$\begin{cases} S_3 = mS_2 - nS_1 \\ S_4 = mS_3 - nS_2 \end{cases} \implies \begin{matrix} 16 = 7m - 3n \\ 42 = 16m - 7n \end{matrix} \implies \begin{cases} m = -14 \\ n = -38 \end{cases}$$

It follows that $S_5 = mS_4 - nS_3 = -14 \times 42 - (-38) \times 16 = \boxed{20}$.

Done.

5.4 Special Sequences

Some sequences given in middle and high school competitions may appear differently from any of the previously discussed types. However, these sequences usually can be transformed to one of these basic types. Let's review a couple of such examples.

Example 5.4.1

Compute
$$S_n = \frac{2}{2} + \frac{3}{2^2} + \frac{4}{2^3} + \cdots + \frac{n+1}{2^n}$$

Solution

This sequence is neither arithmetic nor geometric. However, the denominator of each term forms a geometric sequence and the numerator forms an arithmetic sequence.

Multiplying both sides by 1/2 gives
$$\frac{1}{2} \times S_n = \frac{2}{2^2} + \frac{3}{2^3} + \cdots + \frac{n}{2^n} + \frac{n+1}{2^{n+1}}$$

Subtracting this from the original relation yields:

$$\frac{1}{2} \times S_n = \frac{2}{2} + \left(\frac{1}{2^2} + \frac{1}{2^3} + \cdots + \frac{1}{2^n}\right) - \frac{n+1}{2^{n+1}}$$
$$= 1 + \frac{1}{2^2} \times \frac{1 - 1/2^{n-1}}{1 - 1/2} - \frac{n+1}{2^{n+1}}$$
$$= \frac{3}{2} - \frac{n+3}{2^{n+1}}$$
$$\therefore \quad S_n = \boxed{3 - \frac{n+3}{2^n}}$$

Done.

Example 5.4.2

Suppose sequence $\{a_n\}$ satisfies $a_1 = 0$, $a_2 = 1$, $a_3 = 9$, and

$$S_n^2 S_{n-2} = 10 S_{n-1}^3$$

for $n > 3$ where S_n is the sum of the first n terms of this sequence. Find a_n when $n \geq 3$.

(Ref 2008 Liao Ning)

Similar to tackling high degree equations, substitution also plays an important role in solving irregular sequences.

Solution

Clearly, no S_n will equal 0. Hence dividing both sides of the given recursion by $S_{n-1}^2 S_{n-2}$ leads to

$$\left(\frac{S_n}{S_{n-1}}\right)^2 = 10 \times \left(\frac{S_{n-1}}{S_{n-2}}\right)$$

Let $T_n = \frac{S_n}{S_{n-1}}$, $n \geq 3$. Then $T_3 = \frac{S_3}{S_2} = \frac{0+1+9}{0+1} = 10$ and

$$T_n = 10^{\frac{1}{2}} \times T_{n-1}^{\frac{1}{2}}$$

$$T_{n-1}^{\frac{1}{2}} = 10^{\frac{1}{2^2}} \times T_{n-2}^{\frac{1}{2^2}}$$
$$T_{n-2}^{\frac{1}{2^2}} = 10^{\frac{1}{2^3}} \times T_{n-3}^{\frac{1}{2^3}}$$
$$\cdots$$
$$T_4^{\frac{1}{2^{n+4}}} = 10^{\frac{1}{2^{n+3}}} \times T_3^{\frac{1}{2^{n+3}}}$$

Multiplying these relations and noticing $T_3 = 10$ yields:
$$T_n = 10^{\frac{1}{2}+\frac{1}{2^2}+\cdots+\frac{1}{2^{n+3}}} \times 10^{\frac{1}{2^{n+3}}} = 10$$

This means $S_n/S_{n-1} = T_n = 10$ is a constant which says $\{S_n\}$ is a geometric sequence whose common ratio is 10 and $S_3 = 10$. Hence, by *(5.10)* on *page 47*, we have
$$S_n = 10 \times 10^{n-3} = 10^{n-2} \qquad (n \geq 3)$$

Finally by *(5.1)* on *page 43*, we have
$$a_n = S_n - S_{n-1} = 10^{n-2} - 10^{n-3} = \boxed{9 \times 10^{n-3}} \qquad (n \geq 3)$$

Done.

5.5 Practice

Practice 1

Let S_n be the sum of first n terms of an arithmetic sequence. If $S_n = 30$ and $S_{2n} = 100$, compute S_{3n}.

(Ref 2008 Liao Ning)

Chapter 5: Sequence

Practice 2

Let S_n be the sum of the first n terms in geometric sequence $\{a_n\}$. If all a_n are real numbers and $S_{10} = 10$, and $S_{30} = 70$, compute S_{40}.

(Ref 1998 China)

Practice 3

Expanding
$$\left(\sqrt{x} + \frac{1}{2\sqrt[4]{x}}\right)^n$$
and arranging all the terms in descending order of x's power, if the coefficients of the first three terms form an arithmetic sequence, how many terms with integer power of x are there?

(Ref 2002 China)

Practice 4

Is it possible for a geometric sequence to contain three distinct prime numbers?

Practice 5

Is it possible to construct 12 geometric sequences to contain all the prime numbers between 1 and 100?

(Ref 1995 Russia)

Chapter 5: Sequence

Practice 6

In a sports contest, there were m medals awarded on n successive days ($n > 1$). On the first day, one medal and $1/7$ of the remaining $m - 1$ medals were awarded. On the second day, two medals and $1/7$ of the now remaining medals were awarded; and so on. On the nth and last day, the remaining n medals were awarded. How many days did the contest last, and how many medals were awarded altogether?

(Ref 1967 IMO)

Practice 7

Solve $\{L_n\}$ which is defined as $F_1 = 1, F_2 = 3$ and $F_{n+1} = F_n + F_{n-1}, (n = 2, 3, 4, \cdots)$

Practice 8

Find an expression for x_n if sequence $\{x_n\}$ satisfies $x_1 = 2$, $x_2 = 3$, and

$$\begin{cases} x_{2k+1} = x_{2k} + x_{2k-1} & (k \geq 1) \\ x_{2k} = x_{2k-1} + 2x_{2k-2} & (k \geq 2) \end{cases}$$

(Ref 1983 Australia)

Practice 9

Suppose α and β be two real roots of $x^2 - px + q = 0$ where p and $q \neq 0$ are two real numbers. Let sequence $\{a_n\}$ satisfies $a_1 = p$, $a_2 = p^2 - q$, and $a_n = pa_{n-1} - qa_{n-2}$ for $n > 2$.

i) Express a_n using α and β.

ii) If $p = 1$ and $q = \frac{1}{4}$, find the sum of first n terms of $\{a_n\}$.

(Ref 2009 China)

Practice 10

Suppose sequence $\{F_n\}$ is defined as
$$F_n = \frac{1}{\sqrt{5}}\left[\left(\frac{1+\sqrt{5}}{2}\right)^n - \left(\frac{1-\sqrt{5}}{2}\right)^n\right]$$
for all $n \in \mathbb{N}$. Let
$$S_n = C_n^1 \cdot F_1 + C_n^2 \cdot F_2 + \cdots + C_n^n \cdot F_n.$$
Find all positive integer n such that S_n is divisible by 8.

Chapter 5: Sequence

Practice 11

If sequence $\{a_n\}$ has no zero term and satisfies that, for any $n \in \mathbb{N}$,

$$(a_1 + a_2 + \cdots + a_n)^2 = a_1^3 + a_2^3 + \cdots + a_n^3$$

i) Find all qualifying sequences $\{a_1, a_2, a_3\}$ when $n = 3$.

ii) Is there an infinite sequence $\{a_n\}$ such that $a_{2013} = -2012$? If yes, give its general formula of a_n. If not, explain.

(Ref 2012 China)

Chapter 6

Function

6.1 Domain and Range

Given a function $y = f(x)$, the collection of all permissible values that x can take is called this function's *domain*. Correspondingly, the collection of all the possible values y can have is called this function's *range*. The ability to determine a function's domain and range is required both at school and in math competitions.

Example 6.1.1

What are the domain and range of function $y = |x|$?

Solution

Because x can take all real numbers but y will always be greater or equal to zero, its domain is all real numbers and its range is all non-negative real numbers.

Done.

Chapter 6: Function

Domain and range can be represented using inequality. For example, in the proceeding example, its range can be written as $y \geq 0$. However, more often, domain and range are expressed in set notations. Accordingly, the answer to *Example 6.1.1* can be written as
$$x \in (-\infty, +\infty) \quad \text{and} \quad y \in [0, +\infty)$$

Some special symbols are widely used in set notation. For example:

i) $\mathbb{N}$: natural numbers (positive integers)

ii) $\mathbb{Z}$: integers

iii) $\mathbb{Q}$: rational numbers

iv) $\mathbb{R}$: real numbers

v) $\mathbb{C}$: complex numbers

Using such notation, $f(x) : \mathbb{R} \to \mathbb{R}$ means both the domain and range of function $f(x)$ are real numbers. Similarly, $g(x) : \mathbb{Z} \to \mathbb{R}$ indicates function $g(x)$'s domain is integers and its range is real numbers.

Domain and range related problems in math competitions typically require

i) Solid understanding of certain functions' basic definitions.

ii) Skills in solving inequalities.

Logarithm and exponential functions are popular choices for this type of questions. For instance, the 18th problem in 2014 AMC 12A is an excellent example. It asks for the domain of function
$$f(x) = \log_{\frac{1}{2}}(\log_4(\log_{\frac{1}{4}}(\log_{16}(\log_{\frac{1}{16}} x))))$$

All it takes to solve this problem is a clear understanding of logarithm function's properties and solid inequality solving skills.

Let's review another example.

Example 6.1.2

Find the range of function $y = x + \sqrt{x^2 - 3x + 2}$.
(Ref 2001 China)

Solution

First, we need to guarantee $x^2 - 3x + 2 \geq 0$. This leads to
$$x \geq 2 \quad \text{or} \quad x \leq 1$$
In fact, this is the domain of the given function.

While $x \geq 2$, $x + \sqrt{x^2 - 3x + 2}$ is a monotonically increasing function. It is easy to determine that $y \geq 2$. However, when $x \leq 1$, it is not clear whether the given function is monotonic. Its lower bound can be determined by squeezing:

$$y = x + \sqrt{x^2 - 3x + 2} = x + \sqrt{(1-x)(2-x)}$$
$$\geq x + \sqrt{(1-x)(1-x)} = 1$$

y's minimal value of 1 can be achieved when $x = 1$. However y's upper bound when $x \leq 1$ is difficult to obtain directly.

Given that we know x's range, it is then possible to express x using y and then solve y's range.

$$y = x + \sqrt{x^2 - 3x + 2} \implies x = \frac{y^2 - 2}{2y - 3}$$

Therefore, it must hold that

$$\frac{y^2 - 2}{2y - 3} \leq 1 \quad \text{or} \quad \frac{y^2 - 2}{2y - 3} \geq 2$$

Because we have already handle the 2nd case, $x \geq 2$, above, it is sufficient to just solve
$$\frac{y^2 - 2}{2y - 3} \leq 1$$
Do casework.

Chapter 6: Function

i) If $2y - 3 > 0 \implies y > \frac{3}{2}$, then

$$y^2 - 2 \leq 2y - 3 \implies \text{no solution}$$

ii) If $2y - 3 < 0 \implies y < \frac{3}{2}$, then

$$y^2 - 2 \geq 2y - 3 \implies y < \frac{3}{2}$$

Hence, we conclude $y \in [1, \frac{3}{2}) \cup [2, +\infty)$.

Done.

6.2 Function Properties

Many functions exhibit special properties. At middle school and high school level, those most interesting ones include

- Odd function v.s. even function
- Periodic function
- (Monotonically) increasing or decreasing function

6.2.1 Odd Function v.s. Even Function

A function $f(x)$ is *odd* if, for every x in its domain, it satisfies

$$f(-x) = -f(x) \quad \text{or} \quad f(x) + f(-x) = 0$$

Correspondingly, an *even* function is one that satisfies:

$$f(-x) = f(x) \quad \text{or} \quad f(x) - f(-x) = 0$$

Chapter 6: Function

Geometrically, an odd function has rotational symmetry with respect to the origin. An even function is symmetric with respect to the y-axis.

Examples of odd function include $y = x + 3x^3$, $y = \sin x$, and so on. Examples of even function include $y = x^2$, $y = |x|$, $y = \cos x$, and so on.

Please note that some functions are neither odd nor even. One example is the quadratic function $y = ax^2 + bx + c$ $(a \neq 0)$, unless $b = 0$ in which case it becomes even.

Given a function $f(x)$, we can evaluate $f(-x)$ to determine if it is odd or even.

Example 6.2.1

Is function $f(x) = \lg(x + \sqrt{x^2 + 1})$ an odd or even function?

(Ref 2000 Moscow)

Solution

It is an odd function because

$$f(-x) = \lg(-x + \sqrt{(-x)^2 + 1}) = \lg\left(\frac{1}{x + \sqrt{x^2 + 1}}\right) = -f(x)$$

Alternatively, we can also check

$$f(x) + f(-x) = \lg(x + \sqrt{x^2 + 1}) + \lg(-x + \sqrt{(-x)^2 + 1}) = \lg 1 = 0$$

Done.

6.2.2 Periodic Function

A function $f(x)$ is called *periodic* if there exists a constant $T \neq 0$ such that

$$f(x + T) = f(x)$$

holds for every x in its domain. There exist many periodical functions. For example, all trigonometric functions are periodic because $f(x + 2\pi) = f(x)$ where $f(x)$ is $\sin x$, $\cos x$, and so on. Geometrically, a periodic function exhibits a repeating pattern.

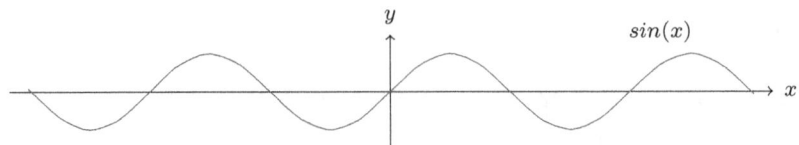

> In order to prove a function is periodic, it is necessary to find a constant $T \neq 0$ such that $f(x + T) = f(x)$.

6.2.3 Monotonic Function

If a function's value is never increasing, it is called monotonically decreasing. Similarly, if a function's value is never decreasing, it is called monotonically increasing. Both of them are called *monotonic* functions. A function can be monotonic over its entire domain or only part of its domain. For example, $y = x^3$ is monotonically increasing over entirely $\mathbb{R}$. However, $y = x^2$ is monotonically decreasing in $(-\infty, 0]$, and monotonically increasing in $[0, +\infty)$.

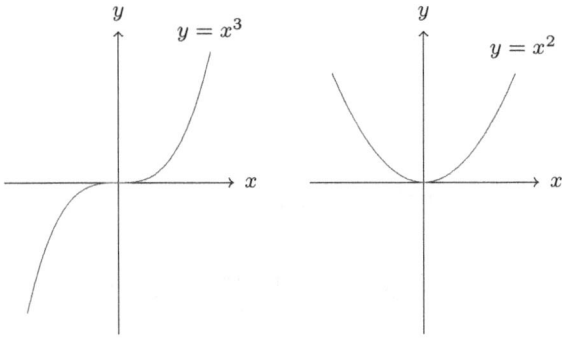

Chapter 6: Function

A general method to determine whether a function is monotonic is to calculate its derivative. A monotonic function's first derivative does not change sign. The converse also holds true. Calculating a function's derivative requires calculus. However, sometimes, it is possible to perform analysis without using calculus.

Example 6.2.2

If real number x satisfies $x^4 - 2x^3 - 7x^2 + 8x + 12 \leq 0$, find the maximum value of $| x + \frac{4}{x} |$.

(Ref China)

When $x > 0$, the minimal value of $(x+\frac{4}{x})$ can be easily computed by applying the AM-GM inequality[1]. The result is 2 which can be achieved when $x = 2$.

However, its maximum value does not exist unless x is bounded within a range. If the domain is a closed range and the function is monotonic within this range, then its maximum value will be achieved at one of the boundaries.

The condition that $x^4 - 2x^3 - 7x^2 + 8x + 12 \leq 0$ will result in a closed range for x. Now, let's show that the function $f(x) = x + \frac{4}{x}$ is monotonically decreasing when $x \leq 2$, and monotonically increasing when $x \geq 2$. This is obvious by observing its function plot. If it is difficult to plot $y = x + \frac{4}{x}$ directly, its property can be analyzed by decomposing it into the sum of two simpler functions $y = x$ and $y = \frac{4}{x}$.

---

[1]Given two non-negative real numbers a and b, it is always true that their arithmetic mean (AM) is not less that their geometric mean (GM):

$$\frac{a+b}{2} \geq \sqrt{ab}$$

Chapter 6: Function

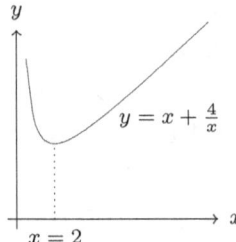
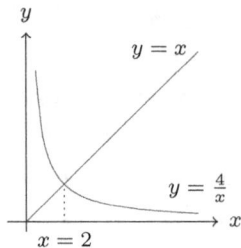

Function $y = x$ is monotonically increasing at a constant slope of 1. Meanwhile, function $y = \frac{4}{x}$ is monotonically decreasing. The absolute value of its slope is greater than 1 when $x < 2$, and less than 1 when $x > 2$. Therefore, when these two functions are added, function $y = \frac{4}{x}$ dominates before $x = 2$ and $y = x$ dominates afterwards. This means $y = x + \frac{4}{x}$ monotonically decreases when $x < 2$, and monotonically increases when $x > 2$.

Let's solve *Example 6.2.2* now.

Solution

Because $x^4 - 2x^3 - 7x^2 + 8x + 12 = (x+1)(x+2)(x-2)(x-3) \leq 0$, the solution is[2]
$$x \in [-2, -1] \cup [2, 3]$$
Because $f(x) = x + \frac{4}{x}$ is an odd function and its value is negative when $x < 0$, the value of $|x + \frac{4}{x}|$ when $x \in [-2, -1]$ is equivalent to the value of $(x + \frac{4}{x})$ when $x \in [1, 2]$. Hence, the original problem is equivalent to finding the maximum value of $(x + \frac{4}{x})$ when $x \in [1, 2] \cup [2, 3] = [1, 3]$.

As function $f(x) = (x + \frac{4}{x})$ is monotonically decreasing when $x < 2$, and monotonically increasing when $x > 2$, its maximum value must be between its left boundary, $f(1) = 5$, and its right boundary, $f(3) = 3\frac{3}{4}$. So the answer is $\boxed{5}$.

Done.

---

[2] See practice problem in *Chapter 3*.

Chapter 6: Function

6.3 Function Equation

While function domain and range related problems often appear in school tests, function equation related problems almost exclusively appear in math competitions. There are two basic types of such problems:

i) Compute the value of a function at specific points.

ii) Find the exact form of a function.

Usually, the 2nd type of problems is much harder. In fact, the vast majority of such problems are seen in USAMO and IMO.

Let's review a couple of examples in this section so readers can get familiar with the basic solving techniques.

Example 6.3.1

For any real number $x \neq 0$, function $f(x)$ satisfies the following relation
$$f(x) + 2f\left(\frac{1}{x}\right) = 3,$$
what are the values of $f(1)$ and $f(2)$?

The basic techniques for solving equation function related problems are

i) Set special values to x.

ii) Treat $f(x)$ as a regular variable and then solve a system of equations.

Solution

Setting $x = 1$ leads to
$$f(1) + 2f(1) = 3 \implies f(1) = \boxed{1}$$

Chapter 6: Function

Setting $x = 2$ yields

$$f(2) + 2f\left(\frac{1}{2}\right) = 3 \qquad (6.1)$$

Setting $x = \frac{1}{2}$ yields

$$f\left(\frac{1}{2}\right) + 2f(2) = 3 \qquad (6.2)$$

Now there are two functions *(6.1)*, *(6.2)* and two variables $f(2)$, $f(\frac{1}{2})$, it will be possible to solve both $f(2)$ and $f(\frac{1}{2})$.

$$2 \times (6.2) - (6.1) \implies 3f(2) = 3 \implies f(2) = \boxed{1}$$

Done.

The next question is that is it possible to determine the exact form of $f(x)$? In this particular case, this task is relatively easy.

Given

$$f(x) + 2f\left(\frac{1}{x}\right) = 3 \qquad (6.3)$$

Replacing x with $\frac{1}{x}$ leads to

$$f\left(\frac{1}{x}\right) + 2f(x) = 3 \qquad (6.4)$$

Solving *(6.3)* and *(6.4)* gives

$$f(x) = \boxed{1}$$

This means that the desired function is a constant.

Here is another example.

Example 6.3.2

Let function $f(x) : \mathbb{R} \to \mathbb{R}$ satisfies $f(0) = 1$ and, for any $x, y \in \mathbb{R}$, $f(x - y) = f(x) - y(2x - y + 1)$. Solve $f(x)$.

Solution

Letting $x = y$ yields $f(x - x) = f(x) - x(2x - x + 1)$. Then, by the given condition, $f(0) = 1$, we have

$$f(x) - x(2x - x + 1) = 1 \implies f(x) = x^2 + x + 1$$

Done.

Both *Example (6.3.1)* and *(6.3.2)* are basic. Some practice problems involve a few advanced solving techniques with detailed explanation given in solutions.

6.4 Practice

Practice 1

Find the range of function $f(x) = 3^{-|\log_2 x|} - 4|x - 1|$.
(Ref 2012 Hope)

Practice 2

Find the minimal value of $y = \sqrt{x^2 + 2x + 5} + \sqrt{x^2 - 4x + 5}$.

Chapter 6: Function

Practice 3

Let $f(x) = x + \frac{4}{x}$. Find the region where the function $f(f(x))$ increase monotonically.

Practice 4

If for any non-negative real numbers x and y, function $f(x)$ satisfies the properties that $f(x) \geq 0$, $f(1) \neq 0$, and $f(x+y^2) = f(x) + 2[f(y)]^2$, compute the value of $f(2 + \sqrt{3})$.

(Ref 2001 Beijing)

Practice 5

Solve

$$\left| \frac{1}{\log_{\frac{1}{2}} x} + 2 \right| > \frac{3}{2}$$

(Ref 2001 China)

Practice 6

If the minimal and maximum values of function

$$f(x) = -\frac{1}{2}x^2 + \frac{13}{2}$$

in the domain $[a, b]$ are $2a$ and $2b$, respectively, determine the values of a and b.

(Ref 2000 China)

Practice 7

Find the function $f(x)$ such that $f(0) = 1$, $f(\frac{\pi}{2}) = 2$, and for any $x, y \in \mathbb{R}$,

$$f(x+y) + f(x-y) = 2f(x)\cos y$$

Practice 8

Let the domain of function $f(n)$ be $\mathbb{N}$, $f(1) = 1$, and for any integer $n \geq 2$,
$$f(n) = f(n-1) + 2^{n-1}$$

Determine $f(n)$.

Practice 9

Let the domain of function $f(n)$ be $\mathbb{N}$, $f(1) = 1$, and for any $m, n \in \mathbb{N}$,
$$f(m+n) = f(m) + f(n) + mn$$

Determine $f(n)$.

Practice 10

Let $f(x) : \mathbb{R} \to \mathbb{R}$. For any $x \in \mathbb{R}$, it always hold that $f(x+3) \leq f(x) + 3$ and $f(x+2) \geq f(x) + 2$. Define $g(x) = f(x) - x$.

i) Prove that $g(x)$ is periodical.

ii) If $f(998) = 1002$, find the value of $f(2000)$.

(Ref 2000 Beijing)

Chapter 6: Function

Practice 11

Find all functions $f : \mathbb{Q} \to \mathbb{Q}$ such that the Cauchy equation
$$f(x+y) = f(x) + f(y)$$
holds for all $x, q \in \mathbb{Q}$.

Practice 12

Solve the following system in integers:
$$\begin{cases} x_1 + x_2 + \cdots + x_n = n \\ x_1^2 + x_2^2 + \cdots + x_n^2 = n \\ \cdots \\ x_1^n + x_2^n + \cdots + x_n^n = n \end{cases}$$

Practice 13

Let x, y, and z be all in $(0, 1)$. Prove
$$x(1-y) + y(1-z) + z(1-x) < 1$$

Practice 14

Show that
$$\frac{(x+a)(x+b)}{(c-a)(c-b)} + \frac{(x+b)(x+c)}{(a-b)(a-c)} + \frac{(x+c)(x+a)}{(b-c)(b-a)} = 1$$

Chapter 7

Solutions

Chapter 7: Solutions

7.1 Introduction

This section is intentionally left blank.

So section numbers of solutions and practices can match.

Chapter 7: Solutions

7.2 Vieta Theorem

Practice 1

Let a and b be the two roots of $x^2 + x + 1 = 0$. Evaluate $a^2 + b^2$. Try to find at least three different solutions without solving the equation directly.

Solution 1

$$a^2 + b^2 = (a+b)^2 - 2ab = (-1)^2 - 2 \times 1 = \boxed{-1}$$

Solution 2

Because a and b are the two roots of $x^2 + x + 1 = 0$, we have
$$a^2 + b^2 = (-a-1) + (-b-1) = -(a+b) - 2 = 1 - 2 = \boxed{-1}$$

Solution 3

Let $y_n = a^n + b^n$, then $y_{n+2} + y_{n+1} + y_n = 0$
$$y_{n+2} = -y_{n+1} - y_n$$
$$\therefore \quad y_0 = 2$$
$$y_1 = -1$$
$$y_2 = -(-1) - 2 = \boxed{-1}$$

Practice 2

Let a and b be the two roots of $x^2 + x + 1 = 0$. Evaluate
$$\frac{1}{a} + \frac{1}{b}$$

Try to find at least three different solutions without solving the equation directly.

Chapter 7: Solutions

Solution 1

$$\frac{1}{a} + \frac{1}{b} = \frac{a+b}{ab} = \frac{-1}{1} = \boxed{-1}$$

Solution 2

Let $y_n = a^n + b^n$, then $y_{n+2} + y_{n+1} + y_n = 0$ or $y_n = -y_{n+2} - y_{n+1}$.

It is clear that $y_0 = a^0 + b^0 = 1 + 1 = 2$ and $y_1 = a + b = -1$.

$$\therefore \quad \frac{1}{a} + \frac{1}{b} = y_{-1} = -y_1 - y_0 = 1 - 2 = \boxed{-1}$$

Solution 3

Because a and b are roots of $x^2 + x + 1 = 0$, it must be true that $\frac{1}{a}$ and $\frac{1}{b}$ are the two roots of

$$\left(\frac{1}{x}\right)^2 + \left(\frac{1}{x}\right) + 1 = 0 \quad \text{or} \quad x^2 + x + 1 = 0$$

Applying Vieta's theorem on this equation gives $\frac{1}{a} + \frac{1}{b} = \boxed{-1}$.

Practice 3

If $m^2 = m + 1$, $n^2 - n = 1$ and $m \neq n$, evaluate:

(i) $m^5 + n^5$ and (ii) $\dfrac{1}{m^5} + \dfrac{1}{n^5}$

Because $m \neq n$, we find m and n are two roots of equation

$$x^2 - x - 1 = 0$$

Let $y_k = m^k + n^k$. Then it must hold $y_{k+2} - y_{k+1} - y_k = 0$, or

$$y_{k+2} = y_{k+1} + y_k$$

Chapter 7: Solutions

i) It follows that:

$$y_0 = m^0 + n^0 = 2 \quad \text{by simple math}$$
$$y_1 = m^1 + n^1 = 1 \quad \text{by Vieta's theorem}$$
$$y_2 = y_1 + y_0 = 3$$
$$y_3 = y_2 + y_1 = 4$$
$$y_4 = y_3 + y_2 = 7$$
$$y_5 = y_4 + y_3 = \boxed{11}$$

ii) The recursion can be written as $y_k = y_{k+2} - y_{k+1}$. Therefore

$$y_{-1} = y_1 - y_0 = -1$$
$$y_{-2} = y_0 - y_{-1} = 3$$
$$y_{-3} = y_{-1} - y_{-2} = -4$$
$$y_{-4} = y_{-2} - y_{-3} = 7$$
$$y_{-5} = y_{-3} - y_{-4} = \boxed{-11}$$

Practice 4

Let a and b be the two roots of $x^2 + x - 1 = 0$. Evaluate $a - b$.

> **Tip:** *Asymmetric expression may lead to multiple answers. Do not forget $\pm$ signs when needed.*

$$a - b = \pm\sqrt{(a-b)^2}$$
$$= \pm\sqrt{(a+b)^2 - 4ab}$$
$$= \pm\sqrt{(-1)^2 - 4 \times (-1)} = \boxed{\pm\sqrt{5}}$$

Chapter 7: Solutions

Practice 5

If $x^2 + 11x + 16 = 0$, $y^2 + 11y + 16 = 0$, and $x \neq y$, what is the value of
$$\sqrt{\frac{x}{y}} - \sqrt{\frac{y}{x}}$$

Clearly, x and y are the two roots of equation $t^2 + 11t + 16 = 0$.

$$\therefore \sqrt{\frac{x}{y}} - \sqrt{\frac{y}{x}} = \frac{y - x}{\sqrt{xy}}$$
$$= \frac{\pm\sqrt{(x+y)^2 - 4xy}}{\sqrt{xy}}$$
$$= \frac{\pm\sqrt{11^2 - 4 \times 16}}{\sqrt{16}}$$
$$= \boxed{\pm \frac{\sqrt{57}}{4}}$$

Practice 6

Let x_1 and x_2 be two roots of $x^2 - x - 1 = 0$. Find the value of
$$2x_1^5 + 5x_2^3$$

Because $x_{1,2}$ are roots of $x^2 - x - 1 = 0$, we have $x_1^2 = x_1 + 1$ and $x_2^2 = x_2 + 1$. Meanwhile, by Vieta's theorem, we find $x_1 + x_2 = 1$.

$$\therefore \quad 2x_1^5 + 5x_2^3 = 2x_1(x_1 + 1)^2 + 5x_2(x_2 + 1)$$
$$= 2x_1(x_1^2 + 2x_1 + 1) + (5x_2^2 + 5x_2)$$
$$= 2x_1((x_1 + 1) + 2x_1 + 1) + (5(x_2 + 1) + 5x_2)$$
$$= 6x_1^2 + 4x_1 + 10x_2 + 5$$
$$= 6(x_1 + 1) + 4x_1 + 10x_2 + 5$$
$$= 10(x_1 + x_2) + 11 = \boxed{21}$$

Chapter 7: Solutions

Practice 7

If the difference of the two roots of the equation $x^2 + 6x + k = 0$ is 2, what is the value of k?

Let the two roots be x_1 and x_2. Then

$$|x_1 - x_2| = 2 \implies 4 = (x_1 - x_2)^2 = (x_1 + x_2)^2 - 4x_1 x_2 = 6^2 - 4k$$

$$\therefore \quad k = 8$$

Practice 8

If one root of the equation $x^2 - 6x + m^2 - 2m + 5 = 0$ is 2. Find the value of the other root and m.

Let the other root be a, then $a + 2 = 6 \implies a = \boxed{4}$.

It follows that $4 \times 2 = m^2 - 2m + 5 \implies m_{1,2} = \boxed{3, -1}$.

Practice 9

Three of the roots of $x^4 + ax^2 + bx + c = 0$ are 2, -3, and 5. Find the value of $a + b + c$.

The coefficient of x^3 is 0, so the sum of all the four roots is 0. This implies the fourth root must be -4.

It is then possible to apply Vieta's formula to compute a, b, and c before summing them up. However there is a quicker way to get the result. Because the four roots are 2, -3, 5, and -4, therefore the following relation must hold:

$$x^4 + ax^2 + bx + c = (x - 2)(x + 3)(x - 5)(x + 4)$$

Chapter 7: Solutions

This is an identity which will hold for any value of x. Let's set $x = 1$:

$$1 + a + b + c = (1-2)(1+3)(1-5)(1+4) = 80 \implies a+b+c = \boxed{79}$$

Practice 10

Let x_1 and x_2 be the two roots of $x^2 - 3mx + 2(m-1) = 0$. If $\frac{1}{x_1} + \frac{1}{x_2} = \frac{3}{4}$, what is the value of m?

By Vieta's theorem,

$$\frac{1}{x_1} + \frac{1}{x_2} = \frac{x_1 + x_2}{x_1 x_2} = \frac{3m}{2(m-1)} = \frac{3}{4} \implies m = \boxed{-1}$$

Practice 11

Let x_1 and x_2 be the two real roots of the equation

$$x^2 - 2(k+1)x + k^2 + 2 = 0$$

If $(x_1 + 1)(x_2 + 1) = 8$, find the value of k

By Vieta's theorem: $x_1 + x_2 = 2(k+1)$ and $x_1 x_2 = k^2 + 2$. Therefore

$$8 = (x_1 + 1)(x_2 + 1) = x_1 x_2 + (x_1 + x_2) + 1 = k^2 + 2 + 2(k+1) + 1$$

or

$$k^2 + 2k - 3 = 0 \implies k_{1,2} = -3, 1$$

Because the original equation has two real roots, we must ensure **its determinant is positive**. This will eliminate $k = -3$. Hence the final answer is $k = \boxed{1}$.

Practice 12

Let α_n and β_n be two roots of equation $x^2 + (2n+1)x + n^2 = 0$ where n is a positive integer. Evaluate the following expression

$$\frac{1}{(\alpha_3+1)(\beta_3+1)} + \frac{1}{(\alpha_4+1)(\beta_4+1)} + \cdots + \frac{1}{(\alpha_{20}+1)(\beta_{20}+1)}$$

(Ref 1983 Shanghai)

By Vieta's theorem:
$$\alpha_n + \beta_n = -(2n+1) \quad \text{and} \quad \alpha_n \cdot \beta_n = n^2$$

Therefore
$$(\alpha_n+1)(\beta_n+1) = \alpha_n \cdot \beta_n + (\alpha_n+\beta_n) + 1 = n^2 - (2n+1) + 1 = (n-2)n$$

It follows that
$$\frac{1}{(\alpha_3+1)(\beta_3+1)} + \frac{1}{(\alpha_4+1)(\beta_4+1)} + \cdots + \frac{1}{(\alpha_{20}+1)(\beta_{20}+1)}$$
$$= \frac{1}{1\times 3} + \frac{1}{2\times 4} + \frac{1}{3\times 5} + \cdots + \frac{1}{18\times 20}$$
$$= \frac{1}{2}\left(\frac{1}{1} - \frac{1}{3}\right) + \frac{1}{2}\left(\frac{1}{2} - \frac{1}{4}\right) + \frac{1}{2}\left(\frac{1}{3} - \frac{1}{5}\right) + \cdots + \frac{1}{2}\left(\frac{1}{18} - \frac{1}{20}\right)$$
$$= \frac{1}{2}\left(\frac{1}{1} + \frac{1}{2} - \frac{1}{19} - \frac{1}{20}\right)$$
$$= \boxed{\frac{531}{760}}$$

Practice 13

If $a \neq 0$ and $\frac{1}{4}(b-c)^2 = (a-b)(c-a)$, compute $\frac{b+c}{a}$.

(Ref 1999 China)

Chapter 7: Solutions

The given condition is equivalent to $(b-c)^2 - 4(a-b)(c-a) = 0$. This implies that the equation
$$(a-b)x^2 - (b-c)x + (c-a) = 0$$
has two equal roots.

Obviously 1 is one root. Therefore both roots are 1. This means
$$\frac{c-a}{a-b} = 1 \times 1 \implies \frac{b+c}{a} = \boxed{2}$$

Practice 14

Let real numbers a, b, c satisfy $a > 0$, $b > 0$, $2c > a+b$, and $c^2 > ab$. Prove
$$c - \sqrt{c^2 - ab} < a < c + \sqrt{c^2 - ab}$$

Let $x_{1,2} = c \pm \sqrt{c^2 - ab}$. Then they are the two roots of
$$x^2 - 2cx + ab = 0$$
$$\begin{aligned}
\therefore \quad (x_1 - a)(x_2 - a) &= x_1 x_2 - (x_1 + x_2)a + a^2 \\
&= ab - 2ca + a^2 \\
&= a(b + a - 2c) \\
&< 0 \quad (\because a > 0, 2c > a+b)
\end{aligned}$$

This means that $(x_1 - a)$ and $(x_2 - a)$ have different signs. Clearly, $x_1 > x_2$. Hence $x_2 < a < x_1$, or
$$c - \sqrt{c^2 - ab} < a < c + \sqrt{c^2 - ab}$$

Practice 15

If both roots of $x^2 + ax + b + 1 = 0$ are positive integers, show that $a^2 + b^2$ cannot be a prime number.

Let the two roots be x_1 and x_2. By Vieta's theorem, we have

$$x_1 + x_2 = -a \quad \text{and} \quad x_1 x_2 = (b+1)$$

Then

$$a^2 + b^2 = (x_1+x_2)^2 + (x_1x_2-1)^2 = x_1^2 + x_2^2 + x_1^2 x_2^2 + 1 = (x_1^2+1)(x_2^2+1)$$

Because both x_1 and x_2 are positive integers, both (x_1^2+1) and (x_2^2+1) are positive integers greater than 1. Hence (a^2+b^2) have two positive divisors both of which are greater than 1. This means it is not prime.

Practice 16

Find integer m such that the equation $x^2 - mx + m + 1 = 0$ has two positive integer roots.

Let x_1 and x_2 be the two roots of $x^2 - mx + m + 1 = 0$. Without loss of generality, let's assume $x_1 \geq x_2$. Then by Vieta's theorem, we have

$$\begin{cases} x_1 + x_2 = m \\ x_1 \cdot x_2 = m + 1 \end{cases}$$

Canceling m leads to

$$x_1 x_2 - x_1 - x_2 = 1 \implies (x_1 - 1)(x_2 - 1) = 2$$

Because both x_1 and x_2 are positive integers, $x_1 \geq x_2$. The above equation can hold if and only if $(x_1 - 1) = 2$ and $(x_2 - 1) = 1$. Or $x_1 = 3$ and $x_2 = 2$. It follows that $m = -(x_1 + x_2) = \boxed{-5}$.

Chapter 7: Solutions

Practice 17

Find the range of real number a if the two roots of $x^2 + 2ax + 6 - a = 0$ satisfy each of the following conditions:

i) two roots are both greater than 1

ii) one root is greater than 1 and the other is less than 1

First, the condition implies the given equation is solvable in real number. Therefore

$$\Delta = (2a)^2 - 4 \times (6-a) = 4a^2 + 4a - 24 \geq 0 \implies a \geq 2 \quad \text{or} \quad a \leq -3$$

i) Let x_1 and x_2 be the two roots of this equation. Both of them are greater than 1 is equivalent to both of $(x_1 - 1)$ and $(x_2 - 1)$ are positive. This means that

$$\begin{cases} (x_1 - 1) + (x_2 - 1) = -2a - 2 > 0 \\ (x_1 - 1)(x_2 - 1) = 6 - a + 2a + 1 > 0 \end{cases}$$

This leads to the following system:

$$\begin{cases} -2a - 2 > 0 \\ 6 - a + 2a + 1 > 0 \\ a \geq 2 \quad \text{or} \quad a \leq -3 \end{cases}$$

Solving this system leads to $\boxed{-7 < a \leq -3}$

ii) Similarly, the condition for one root is greater than 1 and the other is less than 1 is equivalent to the product of $(x_1 - 1)$ and $(x_2 - 1)$ is negative. Or

$$(x_1 - 1)(x_2 - 1) = x_1 x_2 - (x_1 + x_2) + 1 = (6 - a) + 2a + 1 < 0$$

This will give the result as $\boxed{a < -7}$.

Chapter 7: Solutions

Practice 18

Suppose a_1, b_1, c_1, a_2, b_2, and c_2 are all positive real numbers. If both $a_1x^2 + b_1x + c_1 = 0$ and $a_2x^2 + b_2x + c_2 = $ are solvable in real numbers. Show that their roots must be all negative. Furthermore, prove equation $a_1a_2x^2 + b_1b_2x + c_1c_2 = 0$ has two negative real roots too.

Because a_1, b_1 and c_1 are all positive, therefore $-\frac{b_1}{a_1} < 0$ and $\frac{c_1}{a_1} > 0$. By Vieta's theorem, we find the sum of these two roots is negative, but their product is positive. This means both roots are negative. By the same reasoning, the roots of the 2nd equation must be both negative, too.

Because a_1a_2, b_1b_2 and c_1c_2 are all positive, we only need to show that the equation $a_1a_2x^2 + b_1b_2x + c_1c_2 = 0$ is solvable in real numbers in order to show that its two roots are both negative.

$$\Delta = (b_1b_2)^2 - 4(a_1a_2)(c_1c_2) = b_1^2b_2^2 - a_1c_1(4a_2c_2)$$

Because the two original equations are both solvable in real numbers, it must be true that $b_1^2 \geq 4a_1c_1$ and $b_2^2 \geq 4a_2c_2$. This follows that

$$b_1^2b_2^2 - a_1c_1(4a_2c_2) \geq b_1^2(4a_2c_2) - a_1c_1(4a_2c_2) = 4a_2c_2(b_1^2 - a_1c_1) \geq 0$$

Practice 19

Find the sum of all possible integer values of a such that the following equation is solvable in integers:

$$(a+1)x^2 - (a^2+1)x + (2a^2 - 6) = 0$$

If $(a+1) = 0$ or $a = -1$, this equation has an integer root -2.

Chapter 7: Solutions

When $a + 1 \neq 0$, this equation is quadratic. Suppose its two roots are x_1 and x_2. Then, by Vieta's formula, we have

$$x_1 + x_2 = \frac{a^2 + 1}{a + 1} = (a - 1) + \frac{2}{a + 1}$$

If both x_1 and x_2 are integers, 2 must be divisible by $(a+1)$. This implies that $a = 1, 0, -2$ or -3.

- $a = 1 \implies 2x^2 - 2x - 4 = 0 \implies x_{1,2} = 2, -1$
- $a = 0 \implies x^2 - x - 6 = 0 \implies x_{1,2} = 3, -2$
- $a = -2 \implies -x^2 - 5x + 2 = 0$, no integer solution
- $a = -3 \implies -2x^2 - 10x + 12 = 0 \implies x_{1,2} = 1, -6$

Therefore the answer is $-1 + 1 + 0 - 3 = \boxed{-3}$.

Practice 20

In $\triangle ABC$, let a, b, and c be the lengths of sides opposite to $\angle A$, $\angle B$ and $\angle C$, respectively. D is a point on side AB satisfying $BC = DC$. If $AD = d$, show that

$$c + d = 2 \cdot b \cdot \cos A \quad \text{and} \quad c \cdot d = b^2 - a^2$$

As shown in the diagram below, applying Law of Cosines on $\triangle ABC$ and $\triangle ADC$:

$$a^2 = b^2 + c^2 - 2 \cdot b \cdot c \cdot \cos A \quad \text{and} \quad a^2 = b^2 + d^2 - 2 \cdot b \cdot d \cdot \cos A$$

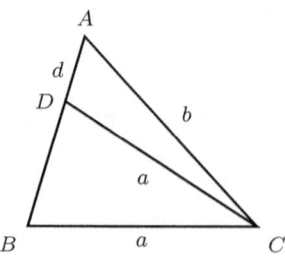

Rearranging these two equations leads to:

$$c^2 - 2 \cdot b \cdot c \cdot \cos A + b^2 - a^2 = 0$$

$$d^2 - 2 \cdot b \cdot d \cdot \cos A + b^2 - a^2 = 0$$

Clearly, we have $c > d$. Therefore they are the two roots of the quadratic equation:

$$x^2 - 2 \cdot b \cdot \cos A \cdot x + (b^2 - a^2) = 0$$

Hence, by Vieta's theorem, the following must hold:

$$c + d = 2 \cdot b \cdot \cos A \quad \text{and} \quad c \cdot d = b^2 - a^2$$

Practice 21

If real numbers m and n satisfy $mn \neq 1$, $19m^2 + 99m + 1 = 0$ and $19 + 99n + n^2 = 0$, what is the value of

$$\frac{mn + 4m + 1}{n}$$

It is obvious that $n \neq 0$ and $m \neq \frac{1}{n}$. Dividing both sides of the 2nd equation by n^2 leads to

$$19\left(\frac{1}{n}\right)^2 + 99\left(\frac{1}{n}\right) + 1 = 0$$

Therefore m and $\frac{1}{n}$ are two distinct roots of the following quadratic equation:

$$19t^2 + 99t + 1 = 0$$

It follows that

$$\frac{mn + 4m + 1}{n} = m + \left(\frac{1}{n}\right) + 4m\left(\frac{1}{n}\right) = \left(-\frac{99}{19}\right) + 4 \times \left(\frac{1}{19}\right) = \boxed{-5}$$

Chapter 7: Solutions

Practice 22

Let α and β be two real roots of $x^4 + k = 3x^2$ and also satisfy $\alpha + \beta = 2$. Find the value of k.

If $\alpha = \beta$, then $\alpha = \beta = 1$, or
$$1^4 + k = 3 \times 1^2 \implies k = 2$$

If $\alpha \neq \beta$, then $\alpha^2 \neq \beta^2$. This is because $\alpha + \beta = 2$ means that $\alpha \neq -\beta$. When $\alpha^2 \neq \beta^2$, α^2 and β^2 must be the two roots of $t^2 + k = 3t$, or
$$t^2 - 3t + k = 0$$
Therefore, by Vieta's theorem:
$$\begin{cases} \alpha^2 + \beta^2 = 3 \\ \alpha^2 \cdot \beta^2 = k \end{cases}$$

$\therefore$ $(\alpha + \beta)^2 = \alpha^2 + \beta^2 + 2\alpha\beta \implies 2^2 = 3 + 2\sqrt{k} \implies k = \dfrac{1}{4}$

When $k = \frac{1}{4}$, the given equation indeed has two roots $1 + \frac{\sqrt{2}}{2}$ and $1 - \frac{\sqrt{2}}{2}$ which satisfy the requirement.

Therefore, k can be either $\boxed{2}$ or $\boxed{\dfrac{1}{4}}$.

Practice 23

Let real numbers a, b, and c satisfy $a + b + c = 2$ and $abc = 4$. Find

i) the minimal value of the largest among a, b, and c.

ii) the minimal value of $|a| + |b| + |c|$.

(Ref 2003 China)

Without loss of generality, suppose $a \geq b \geq c$. This also means that $a > 0$.

Then $b + c = 2 - a$ and $bc = \frac{4}{a}$. Hence, b and c are two real roots of equation
$$x^2 - (2-a)x + \frac{4}{a} = 0$$
Therefore, the discriminant of this equation must be non-negative:
$$(2-a)^2 - 4 \times \frac{4}{a} \geq 0 \implies (a^2+4)(a-4) \geq 0 \implies a \geq \boxed{4}$$
It follows that
$$|a| + |b| + |c| \geq |a| + |b+c| = |a| + |2-a| = 2a - 2 \geq \boxed{6}$$

Practice 24

Compute the value of
$$\sqrt[3]{2 + \frac{10}{3\sqrt{3}}} + \sqrt[3]{2 - \frac{10}{3\sqrt{3}}}$$
and simplify
$$\sqrt[3]{2 + \frac{10}{3\sqrt{3}}} \quad \text{and} \quad \sqrt[3]{2 - \frac{10}{3\sqrt{3}}}$$

Let
$$\alpha = \sqrt[3]{2 + \frac{10}{3\sqrt{3}}} \quad \text{and} \quad \beta = \sqrt[3]{2 - \frac{10}{3\sqrt{3}}}$$

We have
$$\alpha \cdot \beta = \frac{2}{3} \tag{7.1}$$

Suppose
$$\alpha + \beta = r \tag{7.2}$$

Chapter 7: Solutions

Clearly, r is a rational number. Meanwhile,

$$r^3 = \alpha^3 + \beta^3 + 3\alpha\beta(\alpha+\beta) = 4 + 3 \times \frac{2}{3} \times r$$

$$\therefore \quad r^3 - 2r - 4 = 0 \implies (r-2)(r^2 + 2r + 2) = 0$$

This equation has only one rational root, 2, and rational number r obviously is one of its roots. This means that $r = 2$. By (7.2), we have

$$\alpha + \beta = \boxed{2} \tag{7.3}$$

By (7.1) and (7.3), α and β are the two roots of equation

$$t^2 - 2t + \frac{2}{3} = 0$$

Solving this equation and noticing $\alpha > \beta$ yields:

$$\alpha = \boxed{1 + \frac{\sqrt{3}}{3}} \quad \text{and} \quad \beta = \boxed{1 - \frac{\sqrt{3}}{3}}$$

Practice 25

If all coefficients of the polynomial

$$f(x) = a_n x^n + a_{n-1} x^{n-1} + \cdots + a_3 x^3 + x^2 + x + 1$$

are real numbers, prove that its roots cannot be all real.

Let $r_1, r_2, \cdots, r_n$ be the roots of $f(x) = 0$. It is clear that none of them will be zero. Dividing both sides of this equation by x^n:

$$a_n + a_{n-1}\left(\frac{1}{x}\right) + \cdots + \left(\frac{1}{x}\right)^{n-2} + \left(\frac{1}{x}\right)^{n-1} + \left(\frac{1}{x}\right)^n = 0$$

Letting $y = 1/x$ yields

$$g(y) = y^n + y^{n-1} + y^{n-2} + \cdots + a_{n-1} y + a_n = 0$$

Chapter 7: Solutions

Because r_i, $(r = 1, 2, \cdots, n-1, n)$ are the roots of $f(x) = 0$, we find $1/r_i$ must be the roots of $g(y) = 0$. By the Vieta's theorem,

$$\sum_{i=1}^{n}\left(\frac{1}{r_i}\right) = -1 \quad \text{and} \quad \sum_{1 \leq i \leq j \leq n}\left(\frac{1}{r_i} \cdot \frac{1}{r_j}\right) = 1$$

$$\therefore \quad \sum_{i=1}^{n}\left(\frac{1}{r_i}\right)^2 = \left(\sum_{i=1}^{n}\frac{1}{r_i}\right)^2 - 2\sum_{1 \leq i \leq j \leq n}\left(\frac{1}{r_i} \cdot \frac{1}{r_j}\right) = (-1)^2 - 2 \times 1 = -1$$

This means that r_i cannot be all real.

Chapter 7: Solutions

7.3 High Degree Equation

> **Practice 1**

Solve $x^3 - 3x + 2 = 0$.

> ⓘ *Tip: The sum of its coefficients equals zero which implies it is a multiple of $(x-1)$.*

$$x^3 - 3x + 2 = (x-1)^2(x+2) = 0 \implies x_{1,2,3} = \boxed{1, 1, -2}$$

> **Practice 2**

Solve $x^4 - 2x^3 - 7x^2 + 8x + 12 = 0$.

> ⓘ *Tip: If the sum of odd power terms' coefficients equals that of even power terms', then $(x+1)$ must be a divisor.*

Because $4 + (-7) + 12 = (-2) + 8$, it must hold that

$$x+1 \mid x^4 - 2x^3 - 7x^2 + 8x + 12$$

$$\therefore \quad x^4 - 2x^3 - 7x^2 + 8x + 12 = (x+1)(x^3 - 3x^2 - 4x + 12)$$

It follows that

$$\begin{aligned}x^3 - 3x^2 - 4x + 12 &= x^2(x-3) - 4(x-3) \\ &= (x-3)(x^2 - 4) \\ &= (x-3)(x+2)(x-2)\end{aligned}$$

Therefore, we conclude the answers are $-1, 3, -2$, and 2.

Practice 3

Find the range of real number a if equation $\left|\frac{x^2}{x-1}\right| = a$ has exactly two distinct real roots.

(Ref 2006 China)

Clearly, a must be non-negative. Furthermore, if $a = 0$, the given equation has only one root $x = 0$. Hence we conclude $a > 0$.

When $a > 0$, the given equation can be simplified to:

$$\frac{x^2}{x-1} = \pm a \Leftrightarrow x^2 - ax + a = 0 \quad \text{or} \quad x^2 + ax - a = 0$$

Because $a > 0$, the discriminant of the 2nd equation is positive which means it has two distinct real roots. It follows that the 1st equation should be insolvable in real numbers, or

$$a^2 - 4a < 0 \implies a < 4 \quad (\because a > 0)$$

Hence the final answer is $\boxed{0 < a < 4}$.

Practice 4

Solve this equation: $(x^2 - x - 1)^{x+2} = 1$.

Do casework:

i) When $x^2 - x - 1 = 1$, we find $x_{1,2} = -1, 2$.

ii) When $x + 2 = 0$ and $x^2 - x - 1 \neq 0$, we find $x_3 = -2$.

iii) When $x^2 - x - 1 = -1$ and $x + 2$ is even, we find $x_4 = 0$.

Therefore, the given equation has four solutions $\boxed{-2, -1, 0, 2}$.

Chapter 7: Solutions

Practice 5

Solve this system:

$$\begin{cases} \dfrac{4}{3x-2y} + \dfrac{3}{2x-5y} = 10 \\[2mm] \dfrac{5}{3x-2y} - \dfrac{2}{2x-5y} = 1 \end{cases}$$

This system can be solved directly, but is better to be handled by substituting $\frac{1}{3x-2y}$ with u, and $\frac{1}{2x-5y}$ with v.

$$\begin{cases} 4u + 3v = 10 \\ 5u - 2v = 1 \end{cases} \implies \begin{cases} u = 1 \\ v = 2 \end{cases}$$

$$\therefore \begin{cases} \dfrac{1}{3x-2y} = 1 \\[2mm] \dfrac{1}{2x-5y} = 2 \end{cases} \implies \begin{cases} x = \dfrac{4}{11} \\[2mm] y = \dfrac{1}{22} \end{cases}$$

Practice 6

Solve this equation $2x^4 + 3x^3 - 16x^2 + 3x + 2 = 0$.

> **Tip**: Notice that the coefficients in this equation are symmetric.

Because it is obvious that $x \neq 0$, we can divide both sides by x^2 and rearrange these terms:

$$2\left(x^2 + \dfrac{1}{x^2}\right) + 3\left(x + \dfrac{1}{x}\right) - 16 = 0$$

Let $y = x + \frac{1}{x}$, the above equation is equivalent to:

$$2(y^2 - 2) + 3y - 16 = 0 \implies y_{1,2} = \dfrac{5}{2}, -4$$

When $y = \frac{5}{2}$, $x + \frac{1}{x} = \frac{5}{2} \implies x_{1,2} = \frac{1}{2}, 2$.

When $y = -4$, $x + \frac{1}{x} = -4 \implies x_{3,4} = -2 \pm \sqrt{3}$.

Here there are four roots: $\boxed{\frac{1}{2}, 2, -2+\sqrt{3}, -2-\sqrt{3}}$.

Practice 7

Solve this equation $(x-2)(x+1)(x+4)(x+7) = 19$.

One way to solve an equation in the following form
$$(x+a)(x+b)(x+c)(x+d) = e$$
where a, b, c, d, and e are known constants is to pair the four terms into two groups and then to employ the substitution method.

Let's say if we pair $(x+a)$ with $(x+b)$, and $(x+c)$ with $(x+d)$, then the original equation is equivalent to
$$\left(x^2 + (a+b)x + ab\right)\left(x^2 + (c+d)x + cd\right) = e$$

If $a+b = c+d$, then it is possible to transform this 4th degree equation to a quadratic equation by substituting $y = x^2 + (a+b)x$:
$$(y+ab)(y+cd) = e$$

Upon having solved y, it is then possible to solve x.

Solution

Because $-2, 1, 4$, and 7 form an arithmetic sequence, pairing $(x-2)$ with $(x+7)$, and $(x+1)$ with $(x+4)$ will meet the requirement for substitution.

$$\Big((x-2)(x+7)\Big)\Big((x+1)(x+4)\Big) = (x^2+5x-14)(x^2+5x+4) = 19$$

Chapter 7: Solutions

ⓘ *Tip: This equation can already be solved by substituting $(x^2 + 5x)$. Here is an alternative solution.*

Let
$$y = \frac{(x^2 + 5x - 14) + (x^2 + 5x + 5)}{2} = (x^2 + 5x) - 5$$

Then
$$(x^2 + 5x - 14)(x^2 + 5x + 4) = 19$$
$$(y - 9)(y + 9) = 19$$
$$y^2 - 81 = 19$$
$$y_{1,2} = \pm 10$$

When $y = 10$, $x^2 - 5x - 5 = 10 \implies x_{1,2} = \boxed{\dfrac{-5 \pm \sqrt{85}}{2}}$.

When $y = -10$, $x^2 - 5x - 5 = -10 \implies x_{3,4} = \boxed{\dfrac{-5 \pm \sqrt{5}}{2}}$.

Practice 8

Solve equation $(6x + 7)^2(3x + 4)(x + 1) = 6$ in real numbers.

ⓘ *Tip: The left side of this equation is also a product of four terms. Therefore it belongs to the same type as the one in the previous practice. However, it requires one additional step before appropriate pairing can be made.*

The given equation is equivalent to

$$(6x + 7)^2(6x + 8)(6x + 6) = 72 \tag{7.4}$$

Now, after pairing $(6x + 7)^2$, and $(6x + 8)$ with $(6x + 6)$, it is possible to solve this equation by the substitution method as demonstrated in the previous practice. Here, we will illustrate an alternative solution.

Let $a = (6x+7)^2$ and $b = (6x+8)(6x+6)$. Then we have $a + (-b) = 1$. Also, by *(7.4)*, we have $a \times (-b) = -72$. Therefore a and $-b$ are the two roots of the equation

$$t^2 - t - 72 \implies t_{1,2} = -8, 9 \implies a, -b = -8, 9$$

Because $a = (6x+7)^2 \geq 0$, the only possibility is $a = 9$ which means $(a, b) = (9, 8)$. It follows that

$$\begin{cases} (6x+7)^2 &= 9 \\ (6x+8)(6x+6) &= 8 \end{cases} \implies x_{1,2} = \boxed{-\frac{2}{3}, -\frac{5}{3}}$$

Practice 9

If all roots of the equation

$$x^4 - 16x^3 + (81 - 2a)x^2 + (16a - 142)x + (a^2 - 21a + 68) = 0$$

are integers, find the value of a and solve this equation.
(Ref 2009 Jiang Xi)

Re-arranging the given equation to one with respect to a:

$$a^2 - (2x^2 - 16x + 21)a + (x^4 - 16x^3 + 81x^2 - 142x + 68) = 0 \quad (7.5)$$

Because

$$x^4 - 16x^3 + 81x^2 - 142x + 68 = (x^2 - 6x + 4)(x^2 - 10x + 17),$$

and

$$(x^2 - 6x + 4) + (x^2 - 10x + 17) = (2x^2 - 16x + 21)$$

(7.5) can be solved using factorization:

$$a = x^2 - 6x + 4 \quad \text{or} \quad a = x^2 - 10x + 17$$

$$x^2 - 6x + (4 - a) = 0 \quad \text{or} \quad x^2 - 10x + (17 - a) = 0$$

Chapter 7: Solutions

Solving these two equations gives

$$x_{1,2} = 3 \pm \sqrt{5+a} \qquad x_{3,4} = 5 \pm \sqrt{8+a}$$

In order for all x to be integers, both $5+a$ and $8+a$ must be perfect squares. Let

$$5 + a = m^2 \quad \text{and} \quad 8 + a = n^2$$

where m and n are two positive integers satisfying $n > m$.

Hence $n^2 - m^2 = 3 \implies (n+m)(n-m) = 3$. The only solution is $n = 2$ and $m = 1$. It follows

$$a = \boxed{-4}, x_{1,2,3,4} = \boxed{2,4,3,7}$$

Practice 10

Suppose the graph of $f(x) = x^4 + ax^3 + bx^2 + cd + d$, where a, b, c, d are all real constants, passes through three points $A(2, \frac{1}{2})$, $B(3, \frac{1}{3})$, and $C(4, \frac{1}{4})$. Find the value of $f(1) + f(5)$.

Consider the function $g(x) = xf(x) - 1$ which is a 5-degree polynomial equation. Clearly,

$$g(2) = 2f(2) - 1 = 2 \times \frac{1}{2} - 1 = 0$$

$$g(3) = 3f(3) - 1 = 3 \times \frac{1}{3} - 1 = 0$$

$$g(4) = 4f(4) - 1 = 4 \times \frac{1}{4} - 1 = 0$$

This means $2, 3, 4$ are the three roots of $g(x)$. Hence, $g(x)$ can be written as

$$g(x) = xf(x) - 1 = (x-2)(x-3)(x-4)(x^2 + px + q)$$

Setting $x = 0$ into the above equation leads to

$$g(0) = -1 = (-2) \times (-3) \times (-4) \times q \implies q = \frac{1}{24}$$

Setting $x = 1$ and 5 into the previous equation again, respectively, yields:

$$g(1) = f(1) - 1 = -6 \times (1 + p + q)$$
$$g(5) = 5f(5) - 1 = 6 \times (25 + 5p + q)$$

$\therefore f(1) + f(5)$
$= (-6 \times (1 + p + q) + 1) + \frac{1}{5} \times (6 \times (25 + 5p + q) + 1)$
$= 25 + \frac{1}{5} - \frac{24}{5} q == 25 + \frac{1}{5} - \frac{24}{5} \times \frac{1}{24}$
$= \boxed{25}$

Practice 11

Find a quadratic polynomial $f(x) = x^2 + mx + n$ such that

$$f(a) = bc, \quad f(b) = ca, \quad f(c) = ab$$

where a, b, c are three distinct real numbers.

Consider 3-degree polynomial $g(x) = xf(x) - abc$. It is clear that a, b and c are three roots of $g(x)$. As such, we must have

$$g(x) = (x - a)(x - b)(x - c)$$
$\therefore \quad xf(x) - abc = (x - a)(x - b)(x - c)$
$\implies f(x) = x^2 - (a + b + c)x + (ab + bc + ca)$

Chapter 7: Solutions

Practice 12

Let $f(x) = 2016x - 2015$. Solve this equation

$$\underbrace{f(f(f(\cdots f(x))))}_{2017 \text{ iterations}} = f(x)$$

Because $f(x)$ is a 1st degree polynomial, $f(f(f(\cdots f(x))))$ is a 1st degree polynomial too. This means that the given equation has only one root.

Clearly, $f(x) = x$ satisfies the given equation. Therefore, the solution to $f(x) = x$ also satisfies the given equation. Hence, the desired result is

$$f(x) = x \implies 2016x - 2015 = x \implies x = \boxed{1}$$

Chapter 7: Solutions

7.4 Non-Polynomial Equation

Practice 1

Let positive numbers a, b, c, d, e, f satisfy $\frac{bcdef}{a} = 4$, $\frac{acdef}{b} = 9$, $\frac{abdef}{c} = 16$, $\frac{abcef}{d} = \frac{1}{4}$, $\frac{abcdf}{e} = \frac{1}{9}$, and $\frac{abcde}{f} = \frac{1}{16}$. Compute the value of $(a + c + e) - (b + d + f)$.

Multiplying the six given equations and, also, noticing all variables are positive yield

$$(abcdef)^4 = 1 \implies abcdef = 1$$

Now, consider this new relation together with the 1st given relation, we find

$$4 = \frac{bcdef}{a} = \frac{abcdef}{a^2} = \frac{1}{a^2} \implies a = \frac{1}{2}$$

Similarly, we can conclude

$$b = \frac{1}{3}, \quad c = \frac{1}{4}, \quad d = 2, \quad e = 3, \quad f = 4$$

Hence, we conclude that the desired result is $\boxed{-\frac{31}{12}}$.

Practice 2

If $\frac{ab}{a+b} = \frac{1}{15}$, $\frac{bc}{b+c} = \frac{1}{17}$, $\frac{ca}{c+a} = \frac{1}{16}$, find the value of

$$\frac{abc}{ab + bc + ca}$$

(Ref Tai Yuan)

Chapter 7: Solutions

ⓘ Tip: *The given conditions and target expression are ideal candidates for the flipping method.*

Taking reciprocals of the three given relations:

$$\begin{cases} \dfrac{1}{a}+\dfrac{1}{b} = 15 \\[4pt] \dfrac{1}{b}+\dfrac{1}{c} = 17 \\[4pt] \dfrac{1}{c}+\dfrac{1}{a} = 16 \end{cases}$$

This is a typical rotational symmetric system. Adding them together and dividing both sides by 2 will lead to

$$\frac{1}{a}+\frac{1}{b}+\frac{1}{c} = 24$$

Subtracting each of the given conditions from this equation yields

$$\frac{1}{a}=7,\quad \frac{1}{b}=8,\quad \frac{1}{c}=9$$

Taking reciprocal of the target expression:

$$\frac{1}{a}+\frac{1}{b}+\frac{1}{c} = 7+8+9 = 24 \implies \frac{abc}{ab+bc+ca} = \boxed{\frac{1}{24}}$$

Practice 3

Solve this system:

$$\begin{cases} x_1+x_2 = x_2+x_3 = \cdots = x_{1997}+x_{1998} = x_{1998}+x_{1999} = 1 \\[4pt] x_1+x_2+x_3+\cdots+x_{1998}+x_{1999} = 1999 \end{cases}$$

(Ref 1999 Hua Bei)

It can be concluded that $x_1, x_3, \cdots x_{1999}$ are all equal, and $x_2, x_4, \cdots, x_{1998}$ are all equal. Let

$$x_1 = x_3 = x_5 = \cdots = x_{1999} = A$$

$$x_2 = x_4 = x_6 = \cdots = x_{1998} = B$$

where A and B are to constants. Hence

$$\begin{cases} A + B = 1 \\ 1000A + 999B = 1999 \end{cases} \implies A = 1000, \quad B = -999$$

$\therefore x_1 = x_3 = \cdots x_{1999} = \boxed{1000}, x_2 = x_4 = \cdots = x_{1998} = \boxed{-999}$

Practice 4

Let a, b, and c be three distinct numbers such that

$$\frac{a+b}{a-b} = \frac{b+c}{2(b-c)} = \frac{c+a}{3(c-a)}$$

Prove that $8a + 9b + 5c = 0$.

Let

$$\frac{a+b}{a-b} = \frac{b+c}{2(b-c)} = \frac{c+a}{3(c-a)} = k$$

where k is a constant. Then

$$a + b = k(a-b), \quad b + c = k(2(b-c)), \quad c + a = k(3(c-a))$$

The goal is to make the target expression $8a + 9b + 5c$ a linear combination of the three numerators given in the condition. In order to achieve this, let's solve the following system:

$$\begin{cases} x + z = 8 \\ x + y = 9 \\ y + z = 5 \end{cases} \implies \begin{cases} x = 6 \\ y = 3 \\ z = 2 \end{cases}$$

Chapter 7: Solutions

Therefore,

$$\therefore\ 8a + 9b + 5c$$
$$= 6(a+b) + 3(b+c) + 2(c+a)$$
$$= k\Big(6(a-b) + 6(b-c) + 6(c-a)\Big)$$
$$= 0$$

Practice 5

Solve this system:

$$\begin{cases} |x+y| + |x| = 4 \\ 2|x+y| + 3|x| = 9 \end{cases}$$

When x and y have the same sign (or $xy > 0$), the given system is equivalent to

$$\begin{cases} |x| + |y| + |x| = 4 \\ 2 \times (|x| + |y|) + 3|x| = 9 \end{cases} \Longrightarrow \begin{cases} |x| = 1 \\ |y| = 2 \end{cases}$$

Therefore, in this case, there are two solutions

$$(x, y) = (1, 2), (-1, -2)$$

Where x and y have different signs (or $xy < 0$), we have

$$\begin{cases} |x| - |y| + |x| = 4 \\ 2 \times (|x| - |y|) + 3|x| = 9 \end{cases} \Longrightarrow \text{no solution}$$

or

$$\begin{cases} -|x| + |y| + |x| = 4 \\ 2 \times (-|x| + |y|) + 3|x| = 9 \end{cases} \Longrightarrow \begin{cases} |x| = 1 \\ |y| = 4 \end{cases}$$

Therefore, under this circumstance, there are two solutions

$$(x, y) = (1, -4), (-1, 4)$$

In conclusion, there are totally four solutions:
$$(x,y) = (1,2), (1,-4), (-1,4), (-1,-2)$$

Practice 6

Solve this system:
$$\begin{cases} |x+y| = 1 \\ |x| + 2|y| = 3 \end{cases}$$

When $xy \geq 0$, the given system is equivalent to
$$\begin{cases} |x| + |y| = 1 \\ |x| + 2|y| = 3 \end{cases} \implies \text{no solution}$$

When $xy < 0$, then we have
$$\begin{cases} |x| - |y| = 1 \\ |x| + 2|y| = 3 \end{cases} \implies \begin{cases} |x| = \frac{5}{3} \\ |y| = \frac{2}{3} \end{cases}$$

or
$$\begin{cases} -|x| + |y| = 1 \\ |x| + 2|y| = 3 \end{cases} \implies \begin{cases} |x| = \frac{1}{3} \\ |y| = \frac{4}{3} \end{cases}$$

In conclusion, this system has four solutions in total:
$$(x,y) = (\pm\frac{5}{3}, \mp\frac{2}{3}), (\pm\frac{1}{3}, \mp\frac{4}{3})$$

Chapter 7: Solutions

Practice 7

Let x, y, z be three integers satisfying

$$\begin{cases} |x+y| + |y+z| + |z+x| = 4 \\ |x-y| + |y-z| + |z-x| = 2 \end{cases}$$

Compute $x^2 + y^2 + z^2$.

From the 2nd equation, we find at least one of $|x-y|$, $|y-z|$, and $|z-x|$ must equal 0, but obviously they cannot all be zero. This means either $x = y < z$ or $x < y = z$.

If $x = y < z$, then

$$\begin{cases} 2|x| + 2|x+z| = 4 \\ 2|z-x| = 2 \end{cases} \implies z = x+1 \implies |x| + |2x+1| = 2$$

This will lead to two solutions

$$(x, y, z) = (-1, -1, 0), \left(\frac{1}{3}, \frac{1}{3}, \frac{4}{3}\right)$$

If $x < y = z$, then

$$\begin{cases} 2|x+y| + 2|y| = 4 \\ 2|y-x| = 2 \end{cases} \implies x = y-1 \implies |x| + |2x+1| = 2$$

This will lead to two solutions

$$(x, y, z) = \left(-\frac{4}{3}, -\frac{1}{3}, -\frac{1}{3}\right), (0, 1, 1)$$

Because x, y, z are all integers, therefore the desired answer is $\boxed{2}$.

Chapter 7: Solutions

Practice 8

Let x be a positive number. Denote by $\lfloor x \rfloor$ the integer part of x and by $\{x\}$ the decimal part of x. Find the sum of all positive numbers satisfying $5\{x\} + 0.2\lfloor x \rfloor = 25$.

The given equation is equivalent to

$$\{x\} = \frac{125 - \lfloor x \rfloor}{25}, \quad (0 \leq \{x\} < 1)$$

Therefore, $\lfloor x \rfloor$ can take values in $100 < \lfloor x \rfloor \leq 125$. Accordingly,

$$x = \lfloor x \rfloor + \{x\} = 5 + \frac{24}{25}\lfloor x \rfloor$$

It follows that the desired sum equals

$$\left(5 + \frac{24}{25} \times 101\right) + \left(5 + \frac{24}{25} \times 102\right) + \cdots + \left(5 + \frac{24}{25} \times 125\right)$$
$$= 5 \times 25 + \frac{24}{25} \times (101 + 102 + \cdots + 125)$$
$$= \boxed{2837}$$

Practice 9

If $abc = 1$, solve this equation

$$\frac{2ax}{ab + a + 1} + \frac{2bx}{bc + b + 1} + \frac{2cx}{ca + c + 1} = 1$$

Because $abc = 1$, we can transform the given relations as

$$\frac{2ax}{ab + a + 1} = \frac{2ax}{ab + a + abc} = \frac{2x}{b + 1 + bc}$$

$$\frac{2cx}{ca + c + 1} = \frac{2bcx}{cab + bc + b} = \frac{2bcx}{1 + bc + b}$$

Chapter 7: Solutions

$$\Longrightarrow 1 = \frac{2ax}{ab+a+1} + \frac{2bx}{bc+b+1} + \frac{2cx}{ca+c+1}$$
$$= \frac{2x}{b+1+bc} + \frac{2bx}{bc+b+1} + \frac{2bcx}{1+bc+b}$$
$$= \frac{2x(1+b+bc)}{bc+b+1}$$
$$= 2x$$

$$\therefore \quad x = \boxed{\frac{1}{2}}$$

Practice 10

Solve this system
$$\begin{cases} |x| + y &= 12 \\ x + |y| &= 6 \end{cases}$$

(Ref 2007 China)

If $x \geq 0$, then
$$\begin{cases} x + y &= 12 \\ x + |y| &= 6 \end{cases} \Longrightarrow |y| - y = -6$$
This is impossible because $|y| - y \geq 0$.

If $x < 0$, then
$$\begin{cases} -x + y &= 12 \\ x + |y| &= 6 \end{cases} \Longrightarrow |y| + y = 18 \Longrightarrow y = 9 \Longrightarrow x = -3$$
In conclusion, there is only one solution $(9, -3)$.

Practice 11

Solve this equation in real numbers:

$$\sqrt{x} + \sqrt{y-1} + \sqrt{z-2} = \frac{1}{2} \times (x+y+z)$$

Chapter 7: Solutions

From the given condition, because
$$x+y+z-2\sqrt{x}-2\sqrt{y-1}-2\sqrt{z-2} =$$
$$(x-2\sqrt{x}+1)+((y-1)-2\sqrt{y-1}+1)+((z-2)-2\sqrt{z-2}+1)$$

$$\therefore \quad (\sqrt{x}-1)^2+(\sqrt{y-1}-1)^2+(\sqrt{z-2}-1)^2 = 0$$
$$\implies \sqrt{x}-1 = \sqrt{y-1}-1 = \sqrt{z-2}-1 = 0$$
$$\implies x = 1, y = 2, z = 3$$

Practice 12

Let a, b, and c be the lengths of $\triangle ABC$'s three sides. Compute the area of $\triangle ABC$ if the following relations hold:

$$\frac{2a^2}{1+a^2} = b, \qquad \frac{2b^2}{1+b^2} = c, \qquad \frac{2c^2}{1+c^2} = a$$

The given conditions are equivalent to
$$1+\frac{1}{a^2} = \frac{2}{b}, \qquad 1+\frac{1}{b^2} = \frac{2}{c}, \qquad 1+\frac{1}{c^2} = \frac{2}{a}$$

Therefore
$$1+\frac{1}{a^2}+1+\frac{1}{b^2}+1+\frac{1}{c^2} = \frac{2}{b}+\frac{2}{c}+\frac{2}{a}$$
$$\left(1-\frac{1}{a}\right)^2 + \left(1-\frac{1}{b}\right)^2 + \left(1-\frac{1}{c}\right)^2 = 0$$
$$\therefore \quad a = b = c = 1$$

It follows that the area of $\triangle ABC$ is $\boxed{\dfrac{\sqrt{3}}{4}}$.

Practice 13

Find $\lfloor x \rfloor$ where $x = 1+\frac{1}{\sqrt{2}}+\frac{1}{\sqrt{3}}+\cdots+\frac{1}{\sqrt{10000}}$

Chapter 7: Solutions

First, let's show
$$2(\sqrt{n+1} - \sqrt{n}) < \frac{1}{\sqrt{n}} < 2(\sqrt{n} - \sqrt{n-1}) \qquad (7.6)$$

This holds because
$$\frac{1}{2\sqrt{n}} = \frac{1}{\sqrt{n}+\sqrt{n}} > \frac{1}{\sqrt{n+1}+\sqrt{n}} = \sqrt{n+1} - \sqrt{n}$$
$$\implies 2(\sqrt{n+1} - \sqrt{n}) < \frac{1}{\sqrt{n}}$$

Similarly,
$$\frac{1}{2\sqrt{n}} = \frac{1}{\sqrt{n}+\sqrt{n}} < \frac{1}{\sqrt{n}+\sqrt{n-1}} = \sqrt{n} - \sqrt{n-1}$$
$$\implies \frac{1}{\sqrt{n}} < 2(\sqrt{n} - \sqrt{n-1})$$

This means *(7.6)* holds. Therefore,

$$2 \times (\sqrt{3} - \sqrt{2}) < \frac{1}{\sqrt{2}} < 2 \times (\sqrt{2} - \sqrt{1})$$

$$2 \times (\sqrt{4} - \sqrt{3}) < \frac{1}{\sqrt{3}} < 2 \times (\sqrt{3} - \sqrt{2})$$
$$\cdots$$
$$2 \times (\sqrt{n+1} - \sqrt{n}) < \frac{1}{\sqrt{n}} < 2 \times (\sqrt{n} - \sqrt{n-1})$$

Obviously, we have $1 = 1 = 1$.

Adding these relations together gives the following inequality:
$$2\sqrt{n+1} - 2\sqrt{2} + 1 < S < 2\sqrt{n} - 2 + 1$$

Setting $n = 10000 = 100^2$ gives the answer $\boxed{198}$.

Chapter 7: Solutions

7.5 Sequence

Practice 1

Let S_n be the sum of first n terms of an arithmetic sequence. If $S_n = 30$ and $S_{2n} = 100$, compute S_{3n}.
(Ref 2008 Liao Ning)

Noticing that n, $2n$, and $3n$ form an arithmetic sequence, we claim that S_n, $(S_{2n} - S_n)$, $(S_{3n} - S_{2n})$, $\cdots$ form an arithmetic sequence.

$$\therefore (S_{2n} - S_n) - S_n = (S_{3n} - S_{2n}) - (S_{2n} - S_n) \implies S_{3n} = \boxed{210}$$

This conclusion can be proved as following. Let a_1 and d be this arithmetic sequence's first term and common difference, respectively. Pair corresponding terms with distance being n by *(5.4)* on *page 44*

$$a_{n+1} = a_1 + nd$$
$$a_{n+2} = a_2 + nd$$
$$\cdots$$
$$a_{2n} = a_n + nd$$

The sum of all terms on the left equals

$$a_n + a_{n+1} + \cdots + a2n = S_{2n} - S_n$$

The sum of all terms on the right equals

$$a_1 + a_2 + \cdots + a_n + n^2 d = S_n + n^2 d$$

$$\therefore \quad (S_{2n} - S_n) - (S_n) = n^2 d$$

Similarly, we can show that

$$(S_{3n} - S_{2n}) - (S_{2n} - S_n) = n^2 d$$

Chapter 7: Solutions

Hence, we can conclude that

$$S_n, (S_{2n} - S_n), S_{3n} - S_{2n}$$

form an arithmetic sequence.

This conclusion can also be visualized. Let's observe the two blocks, S_n and $S_{2n} - S_n$ below. Each pair of elements differ by nd. Hence, their sum must differ by $nd \times n = n^2 d$ which is a constant if n is fixed.

This means sums of these blocks S_n, $S_{2n} - S_n$, $S_{3n} - S_{2n}$, $\cdots$ form an arithmetic sequence whose common difference is $n^2 d$.

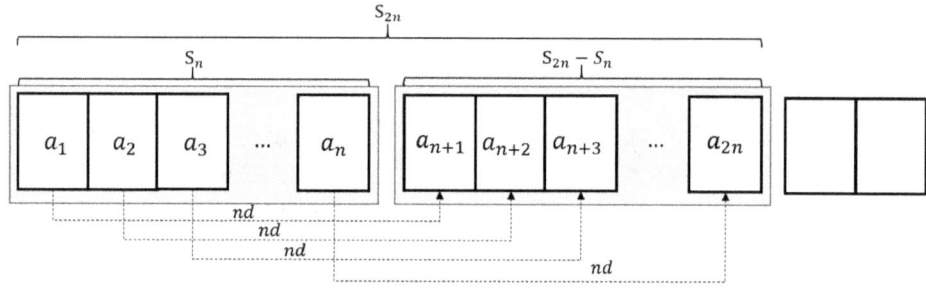

Practice 2

Let S_n be the sum of the first n terms in geometric sequence $\{a_n\}$. If all a_n are real numbers and $S_{10} = 10$, and $S_{30} = 70$, compute S_{40}.

(Ref 1998 China)

Let's first show that S_{10}, $S_{20} - S_{10}$, $S_{30} - S_{20}$, $\cdots$ form a geometric sequence whose initial value is S_{10} and common ratio is r^{10} where r is the common ratio of the original sequence $\{a_n\}$.

$$S_{10} = a_1 \cdot \frac{1 - r^{10}}{1 - r}$$

$$S_{20} - S_{10} = a_1 \cdot \frac{1 - r^{20}}{1 - r} - a_1 \cdot \frac{1 - r^{10}}{1 - r} = a_1 \cdot \frac{1 - r^{10}}{1-} \cdot r^{10}$$

$$S_{30} - S_{20} = a_1 \cdot \frac{1-r^{30}}{1-r} - a_1 \cdot \frac{1-r^{20}}{1-r} = a_1 \cdot \frac{1-r^{10}}{1-r} \cdot r^{20}$$

$$S_{40} - S_{30} = a_1 \cdot \frac{1-r^{40}}{1-r} - a_1 \cdot \frac{1-r^{30}}{1-r} = a_1 \cdot \frac{1-r^{10}}{1-r} \cdot r^{30}$$

Suppose $R = r^{10} \geq 0$, we will then have

$$\begin{aligned} S_{30} &= S_{10} + (S_{20} - S10) + (S_{30} - S_{20}) \\ &= S_{10} + S_{10} \cdot R + S_{10} \cdot R^2 \end{aligned}$$

$\therefore \quad 70 = 10 \times (1 + R + R^2) \implies R = 2$

$\therefore \quad \begin{aligned} S_{40} &= S_{30} + (S_{40} - S_{30}) \\ &= S_{30} + S_{10} \cdot R^3 \\ &= 70 + 10 \times 2^3 \\ &= \boxed{150} \end{aligned}$

Practice 3

Expanding

$$\left(\sqrt{x} + \frac{1}{2\sqrt[4]{x}}\right)^n$$

and arranging all the terms in descending order of x's power, if the coefficients of the first three terms form an arithmetic sequence, how many terms with integer power of x are there?
(Ref 2002 China)

Because

$$\begin{aligned} &\left(\sqrt{x} + \frac{1}{2\sqrt[4]{x}}\right)^n \\ &= \left(x^{\frac{1}{2}} + \frac{1}{2}x^{-\frac{1}{4}}\right)^n \\ &= \sum_{i=0}^{n} C_n^i x^{\frac{n}{2}} \cdot \frac{1}{2^{n-i}} x^{-\frac{n-i}{4}} \\ &= \sum_{i=0}^{n} \frac{1}{2^{n-i}} C_n^i x^{\frac{n+i}{4}} \end{aligned}$$

115

Chapter 7: Solutions

Therefore the bigger the value i is, the bigger the x's power is. This means the three coefficients are

$$\frac{1}{2^{n-n}}C_n^n, \quad \frac{1}{2^{n-(n-1)}}C_n^{n-1}, \quad \frac{1}{2^{n-(n-2)}}C_n^{n-2}$$

Or

$$1, \quad \frac{n}{2}, \quad \frac{1}{4} \times \frac{n(n-1)}{2}$$

As they form an arithmetic sequence, it must hold that

$$2 \times \left(\frac{n}{2}\right) = 1 + \frac{n(n-1)}{8} \implies n = 1, 8$$

Clearly, $x > 1$, hence $n = 8$ is the only possibility. When $n = 8$, every term will be in the following form

$$\frac{1}{2^{8-i}}C_8^i x^{\frac{8+i}{4}}, \quad i = 0, 1, 2, \cdots 8$$

Hence, the power of x will be an integer if and only if $i = 0, 4, 8$ which means the answer is $\boxed{3}$.

Practice 4

Is it possible for a geometric sequence to contain three distinct prime numbers?

The answer is NO. This conclusion can be shown with proof by contradiction.

Suppose it is possible. Let three prime numbers $p_1 < p_2 < p_3$ be three terms of a geometric sequence whose common ratio is r. Then by *(5.10)* on *page 47*, there exist two positive integers m and n such that

$$p_2 = p_1 \cdot r^m \quad \text{and} \quad p_3 = p_2 \cdot r^n$$

Solving r using these two equations, respectively, yields:

$$r = p_1^{-\frac{1}{m}} \cdot p_2^{\frac{1}{m}} = p_2^{-\frac{1}{n}} \cdot p_3^{\frac{1}{n}} \implies p_2^{m+n} = p_1^n \cdot p_3^m$$

This means p_3 divides p_2 which is impossible.

Chapter 7: Solutions

Practice 5

Is it possible to construct 12 geometric sequences to contain all the prime numbers between 1 and 100?
(Ref 1995 Russia)

It is impossible. Because there are 25 prime numbers between 1 and 100. One geometric sequence can hold at most two prime numbers by the result of previous practice. Therefore it is impossible for 12 geometric sequences to have all these 25 prime numbers.

Practice 6

In a sports contest, there were m medals awarded on n successive days ($n > 1$). On the first day, one medal and $1/7$ of the remaining $m-1$ medals were awarded. On the second day, two medals and $1/7$ of the now remaining medals were awarded; and so on. On the nth and last day, the remaining n medals were awarded. How many days did the contest last, and how many medals were awarded altogether?
(Ref 1967 IMO)

Let a_k be the number of medals awarded on the kth day. Then

$$a_1 = 1 + \frac{1}{7} \times (m-1) = \frac{1}{7} \times (m+6)$$
$$a_2 = 2 + \frac{1}{7} \times (m - a_1 - 2)$$
$$\cdots$$
$$a_k = k + \frac{1}{7} \times (m - a_1 - a_2 - \cdots - a_{k-1} - k)$$
$$a_{k+1} = (k+1) + \frac{1}{7} \times (m - a_1 - a_2 - \cdots - a_k - (k+1))$$
$$\cdots$$

Chapter 7: Solutions

Hence the following recursion holds for $k \geq 1$:

$$a_{k+1} - a_k = 1 + \frac{1}{7} \times (-a_k - 1) \implies a_{k+1} = \frac{6}{7}a_k + \frac{6}{7}$$

Suppose

$$a_{k+1} + \alpha = \beta(a_k + \alpha) \implies a_{k+1} = \beta a_k + \alpha(\beta - 1)$$

Comparing the coefficients of these two relations leads to

$$\alpha = -6, \beta = \frac{6}{7} \implies a_{k+1} - 6 = \frac{6}{7} \times (a_k - 6)$$

This means that $\{a_k - 6\}$ forms a geometric sequence whose common ratio is $\frac{6}{7}$ and initial value equals

$$a_1 - 6 = \frac{1}{7} \times (m + 6) - 6 = \frac{1}{7} \times (m - 36)$$

This is followed by

$$a_k - 6 = \frac{1}{7} \times (m - 36) \times \left(\frac{6}{7}\right)^{m-1}$$

Because totally m medals were awarded in n days, we have

$$m = a_1 + a_2 + \cdots + a_n$$
$$= \frac{1}{7} \times (m - 36)\left(1 + \frac{6}{7} + \left(\frac{6}{7}\right)^2 + \cdots + \left(\frac{6}{7}\right)^{n-1}\right) + 6n$$
$$= (m - 36)\left(1 - \left(\frac{6}{7}\right)^n\right) + 6n$$

$$\therefore \quad m = \frac{7^n}{6^{n-1}} \times (n - 6) + 36$$

Because both m and n are natural numbers, 7^n and 6^{n-1} must be co-prime. This means 6^{n-1} must divide $(n - 6)$. However it is obviously that $\mid n - 6 \mid < 6^{n-1}$ for any $n > 1$. This leads to the conclusion that $n - 6 = 0 \implies n = 6$. Accordingly, $m = 36$.

In another word, totally $\boxed{36}$ medals were awarded in $\boxed{6}$ days.

Chapter 7: Solutions

Practice 7

Solve $\{L_n\}$ which is defined as $F_1 = 1, F_2 = 3$ and $F_{n+1} = F_n + F_{n-1}, (n = 2, 3, 4, \cdots)$

This recursion relation is the same as Fibonacci sequence, but the initial values are different. Therefore, the result will be in the same form as that of Fibonacci sequence except the two constant coefficients will be different.

The characteristic equation is $x^2 = x + 1$ whose two roots are $\frac{1 \pm \sqrt{5}}{2}$. Therefore the solution is

$$F_n = C_1 \times \left(\frac{1+\sqrt{5}}{2}\right)^n + C_2 \times \left(\frac{1-\sqrt{5}}{2}\right)^n$$

Setting $F_1 = 1$ and $F_2 = 3$:

$$\begin{cases} 1 = C_1 \times \left(\frac{1+\sqrt{5}}{2}\right)^1 + C_2 \times \left(\frac{1-\sqrt{5}}{2}\right)^1 \\ 3 = C_1 \times \left(\frac{1+\sqrt{5}}{2}\right)^2 + C_2 \times \left(\frac{1-\sqrt{5}}{2}\right)^2 \end{cases} \implies C_1 = C_2 = 1$$

Therefore, we have

$$F_n = \left(\frac{1+\sqrt{5}}{2}\right)^n + \left(\frac{1-\sqrt{5}}{2}\right)^n$$

Practice 8

Find an expression for x_n if sequence $\{x_n\}$ satisfies $x_1 = 2$, $x_2 = 3$, and

$$\begin{cases} x_{2k+1} = x_{2k} + x_{2k-1} & (k \geq 1) \\ x_{2k} = x_{2k-1} + 2x_{2k-2} & (k \geq 2) \end{cases}$$

(Ref 1983 Australia)

Chapter 7: Solutions

> **i** *Tip: Notice that the recursion is given as two separate expressions. In such a case, it will be incorrect to apply the method of characteristic equation directly.*

The first step is to obtain a self-contained recursion. By the 1st recursion $x_{2k+1} = x_{2k} + x_{2k-1}$, we have

$$x_{2k} = x_{2k+1} - x_{2k-1} \implies x_{2k-2} = x_{2k-1} - x_{2k-3}$$

Substituting x_{2k} and x_{2k-2} in the 2nd recursion using these two relations leads to

$$x_{2k+1} - x_{2k-1} = x_{2k-1} + 2(x_{2k-1} - x_{2k-3}) \implies x_{2k+1} = 4x_{2k-1} - 2x_{2k-3}$$

Letting $y_k = x_{2k-1}$ implies

$$y_{k+1} = 4y_k - 2y_{k-1}$$

whose characteristic equation and its roots are

$$t^2 = 4t - 2 \implies t_{1,2} = 2 \pm \sqrt{2}$$

Hence, the solution for y_k is

$$y_k = C_1(2+\sqrt{2})^{k-1} + C_2(2-\sqrt{2})^{k-1}$$

Setting $y_1 = x_1 = 2$ and $y_2 = x_3 = x_2 + x_1 = 5$ to the above solution gives:

$$C_{1,2} = \frac{4 \pm \sqrt{2}}{4}$$

or

$$x_{2k-1} = y_k = \frac{4+\sqrt{2}}{4} \times (2+\sqrt{2})^{k-1} + \frac{4-\sqrt{2}}{4} \times (2-\sqrt{2})^{k-1}$$

It follows that

$$x_{2k} = x_{2k+1} - x_{2k-1}$$
$$= \left(\frac{4+\sqrt{2}}{4} \times (2+\sqrt{2})^k + \frac{4-\sqrt{2}}{4} \times (2-\sqrt{2})^k \right)$$

$$-\left(\frac{4+\sqrt{2}}{4} \times (2+\sqrt{2})^{k-1} + \frac{4-\sqrt{2}}{4} \times (2-\sqrt{2})^{k-1}\right)$$

$$= \frac{2\sqrt{2}+1}{4} \times (2+\sqrt{2})^k - \frac{2\sqrt{2}-1}{4} \times (2-\sqrt{2})^k$$

Therefore the desired answer is

$$x_n = \begin{cases} \frac{2\sqrt{2}+1}{4} \times (2+\sqrt{2})^k - \frac{2\sqrt{2}-1}{4} \times (2-\sqrt{2})^k & (n=2k) \\ \frac{4+\sqrt{2}}{4} \times (2+\sqrt{2})^{k-1} + \frac{4-\sqrt{2}}{4} \times (2-\sqrt{2})^{k-1} & (n=2k-1) \end{cases}$$

where $k = 1, 2, 3, \cdots$.

Practice 9

Suppose α and β be two real roots of $x^2 - px + q = 0$ where p and $q \neq 0$ are two real numbers. Let sequence $\{a_n\}$ satisfies $a_1 = p$, $a_2 = p^2 - q$, and $a_n = pa_{n-1} - qa_{n-2}$ for $n > 2$.

i) Express a_n using α and β.

ii) If $p = 1$ and $q = \frac{1}{4}$, find the sum of first n terms of $\{a_n\}$.

(Ref 2009 China)

Clearly the characteristic equation of a_n is $x^2 = px - q$ whose two roots are α and β. Hence by *Theorem 5.3.1* on *page 51*

If $\alpha \neq \beta$, then
$$a_n = C_1 \alpha^n + C_2 \beta^n$$

Otherwise, if $\alpha = \beta$, then
$$a_n = (C_1 + C_2 n)\alpha^n$$

Chapter 7: Solutions

where C_1 and C_2 are two constants which can be determined by the given initial values $a_1 = p$ and $a_2 = p^2 - q$. The results are

$$a_n = \begin{cases} \dfrac{\alpha^{n+1} - \beta^{n+1}}{\alpha - \beta} & (\alpha \neq \beta) \\ (1+n)\alpha^n & (\alpha = \beta) \end{cases}$$

When $p = 1$ and $q = \frac{1}{4}$, the equation has two equal roots $\alpha = \beta = \frac{1}{2}$. Therefore

$$a_n = (1+n) \times \left(\frac{1}{2}\right)^n = \frac{n+1}{2^n}$$

Let S_n be the sum of first n terms, or

$$S_n = \frac{2}{2} + \frac{3}{2^2} + \frac{4}{2^3} + \cdots + \frac{n+1}{2^n} = \boxed{3 - \frac{n+3}{2^n}}$$

by the result of *Example 5.4.1* on *page 54*.

Practice 10

Suppose sequence $\{F_n\}$ is defined as

$$F_n = \frac{1}{\sqrt{5}}\left[\left(\frac{1+\sqrt{5}}{2}\right)^n - \left(\frac{1-\sqrt{5}}{2}\right)^n\right]$$

for all $n \in \mathbb{N}$. Let

$$S_n = C_n^1 \cdot F_1 + C_n^2 \cdot F_2 + \cdots + C_n^n \cdot F_n.$$

Find all positive integer n such that S_n is divisible by 8.

For convenience, let $\alpha = \frac{1+\sqrt{5}}{2}$ and $\beta = \frac{1-\sqrt{5}}{2}$. Then

$$S_n = \frac{1}{\sqrt{5}}\left[\left(1 + C_n^1 \alpha + C_n^2 \alpha^2 + \cdots + C_n^n \alpha^n\right)\right.$$

$$-\left(1 + C_n^1\beta + C_n^2\beta^2 + \cdots + C_n^n\beta^n\right)\Big]$$
$$= \frac{1}{\sqrt{5}}\Big[(\alpha+1)^n - (\beta+1)^n\Big]$$
$$= \frac{1}{\sqrt{5}}\left[\left(\frac{3+\sqrt{5}}{2}\right)^n - \left(\frac{3-\sqrt{5}}{2}\right)^n\right]$$

Therefore the two roots to sequence $\{S_n\}$'s characteristic equation are $\frac{3\pm\sqrt{5}}{2}$. This means the equation is

$$t^2 - 3t + 1 = 0$$

Accordingly, the recursion is

$$S_{n+2} = 3S_{n+1} - S_n \qquad (n = 1, 2, 3, \cdots) \qquad (7.7)$$

The two initial values are:

$$S_1 = \frac{1}{\sqrt{5}}\left[\left(\frac{3+\sqrt{5}}{3}\right) - \left(\frac{3-\sqrt{5}}{3}\right)\right] = 1$$
$$S_2 = \frac{1}{\sqrt{5}}\left[\left(\frac{3+\sqrt{5}}{3}\right)^2 - \left(\frac{3-\sqrt{5}}{3}\right)^2\right] = 3$$

Because both S_1 and S_2 are integers, by recursion *(7.7)*, all terms in $\{S_n\}$ are integers. Meanwhile it is easy to see that

$$5S_{n+1} = 15S_n - 5S_{n-1} \qquad (n = 2, 3, 4, \cdots) \qquad (7.8)$$

Subtracting *(7.8)* from *(7.7)* gives

$$S_{n+2} = 8S_{n+1} - 16S_n + 5S_{n-1} \qquad (n = 2, 3, 4, \cdots)$$

or

$$S_{n+3} = 8S_{n+2} - 16S_{n+1} + 5S_n \qquad (n = 1, 2, 3, \cdots)$$

Now it is clear from this recursion that S_{n+3} is divisible by 8 if and only if S_n is divisible by 8. Because $S_1 = 1$ and $S_2 = 3$, both are indivisible by 8. But $S_3 = 3 \times 3 - 1 = 8$ is. Therefore, we find S_n is divisible by 8 if and only if $n = \boxed{3k}$ where k is a positive integer.

Chapter 7: Solutions

Practice 11

If sequence $\{a_n\}$ has no zero term and satisfies that, for any $n \in \mathbb{N}$,

$$(a_1 + a_2 + \cdots + a_n)^2 = a_1^3 + a_2^3 + \cdots + a_n^3$$

i) Find all qualifying sequences $\{a_1, a_2, a_3\}$ when $n = 3$.

ii) Is there an infinite sequence $\{a_n\}$ such that $a_{2013} = -2012$? If yes, give its general formula of a_n. If not, explain.

(Ref 2012 China)

i) When $n = 1$, $a_1^2 = a_1^3 \implies a_1 = 1$ ($\because a_i \neq 0$).

When $n = 2$, $(1 + a_2)^2 = 1^3 + a_2^3 \implies a_2 = 2$ or -1.

When $n = 3$, $(1 + a_2 + a_3)^2 = 1^3 + a_2^3 + a_3^3$

a) If $a_2 = 2 \implies a_3 = 3$ or -2.

b) If $a_2 = -1 \implies a_3 = 1$.

Therefore, we conclude there are three possibilities:

$$\{1, 2, 3\} \quad , \quad \{1, 2, -2\} \quad \text{and} \quad \{1, -1, 1\}$$

ii) Let $S_n = a_1 + a_2 + \cdots + a_n$. Then

$$S_n^2 = a_1^3 + a_2^3 + \cdots + a_n^3 \tag{7.9}$$

and

$$(S_n + a_{n+1})^2 = a_1^3 + a_2^3 + \cdots + a_n^3 + a_{n+1}^3 \tag{7.10}$$

Now, *(7.10)* - *(7.9)* yields

$$2S_n = a_{n+1}^2 - a_{n+1}$$

We know that when $n = 1$, $a_1 = 1$. When $n \geq 2$,

$$2a_n = 2(S_n - S_{n-1}) = (a_{n+1}^2 - a_{n+1}) - (a_n^2 - a_n)$$
$$\implies a_{n+1}^2 - a_{n+1} - a_n^2 - a_n = 0$$
$$\implies (a_{n+1} + a_n)(a_{n+1} - a_n - 1) = 0$$

$$\therefore a_{n+1} = -a_n \quad \text{or} \quad a_{n+1} = a_n + 1$$

Setting $a_=1$ and $a_{2013} = -2012$ gives one solution

$$a_n = \begin{cases} n & (1 \leq n \leq 2012) \\ (-1)^n \times 2012 & (n \geq 2013) \end{cases}$$

Chapter 7: Solutions

7.6 Function

Practice 1

Find the range of function $f(x) = 3^{-|\log_2 x|} - 4|x-1|$.
(Ref 2012 Hope)

It is easy to determine that $-4|x-1|$ will reach its maximum value of 0 when $x = 1$.

Meanwhile, exponential function is monotonically increasing when its base is greater than 1. Therefore, $3^{-|\log_2 x|}$ will reach its maximum when $-|\log_2 x|$ is the largest, or $x = 1$.

Hence, the max $f(x) = f(1) = 1$.

Obviously, the minimal value of $f(x)$ is $-\infty$ and this function is continuous on its entire domain which means its range is $\boxed{(-\infty, 1]}$.

Practice 2

Find the minimal value of $y = \sqrt{x^2 + 2x + 5} + \sqrt{x^2 - 4x + 5}$.

This problem can be solved using geometric method. The given equation can be rewritten as

$$y = \sqrt{(x+1)^2 + (0-2)^2} + \sqrt{(x-2)^2 + (0+1)^2}$$

Therefore, finding the minimal value of y is equivalent to find a point on x−axis such that the sum of its distance to point $A(-1, 2)$ and point $B(2, -1)$ minimizes.

Notice that A and B are on different sides of the x−axis. Therefore, the minimal value of the sum is simply the distance between

A and B which is

$$\sqrt{(-1-2)^2 + (2-(-1)^2)} = \boxed{3\sqrt{2}}$$

Practice 3

Let $f(x) = x + \frac{4}{x}$. Find the region where the function $f(f(x))$ increase monotonically.

It is clear that $f(x)$ is an odd function, which means it is symmetric with respect to the origin. Therefore we can first check its property when $x > 0$ and then derive its global property using its symmetric feature.

When $x > 0$, by applying a similar logic in *Example 6.2.2* on *page 67*, we find $f(x)$ is monotonically increasing in $[2, +\infty)$. Therefore, when $x \in \mathbb{R}$, $f(x)$ is monotonically in

$$\mathbb{A} = (-\infty, -2] \cup [2, +\infty) \tag{7.11}$$

Now, let's examine the feature of $f(f(x))$. It can be shown that the desired region is the same as the underlying region, i.e. *(7.11)*. This is because when $x \in [2, +\infty)$, $f(x)$ monotonically increases in $[4, +\infty)$. Because $[4, +\infty)$ is a subset of the $[2, +\infty)$, therefore $f(f(x))$ must be monotonically increasing too. However, when $x \in [0, 2]$, $f(x)$ will be decreasing monotonically in $(\infty, 4]$. As a result, $f(f(x))$ will not be monotonically increasing in this region.

Chapter 7: Solutions

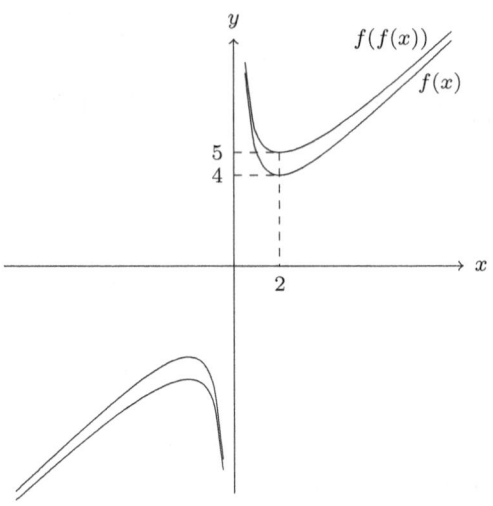

Practice 4

If for any non-negative real numbers x and y, function $f(x)$ satisfies the properties that $f(x) \geq 0$, $f(1) \neq 0$, and $f(x+y^2) = f(x) + 2[f(y)]^2$, compute the value of $f(2 + \sqrt{3})$.

(Ref 2001 Beijing)

Because

$$f(2 + \sqrt{3}) = f(2 + (\sqrt[4]{3})^2) = f(2) + 2[f(\sqrt[4]{3})]^2, \quad (7.12)$$

we will need to compute $f(2)$ and $2[f(\sqrt[4]{3})]^2$.

Setting $x = y = 0$, then

$$f(0 + 0^2) = f(0) + 2[f(0)]^2 \implies f(0) = 0$$

Setting $x = 1, y = 0$, then

$$f(1 + 0^2) = f(1) + 2[f(0)]^2 \implies f(1) = \frac{1}{2}$$

Setting $x = y = 1$, the

$$f(1 + 1^2) = f(1) + 2[f(1)]^2 \implies f(2) = 1 \quad (7.13)$$

Setting $x = 2, y = 1$, then

$$f(2+1^2) = f(2) + 2[f(1)]^2 \implies f(3) = \frac{3}{2}$$

Setting $x = 0, y = \sqrt{3}$, then

$$f(0+(\sqrt{3})^2) = f(0) + 2[f(\sqrt{3})]^2 \implies f(\sqrt{3}) = \frac{\sqrt{3}}{2}$$

Setting $x = 0, y = \sqrt[4]{3}$, then

$$f(0+(\sqrt[4]{3})) = f(0) + 2[f(\sqrt[4]{3})]^2 \implies 2[f(\sqrt[4]{3})]^2 = \frac{\sqrt{3}}{2} \quad (7.14)$$

$$\therefore \quad f(2+\sqrt{3}) = \boxed{1 + \frac{\sqrt{3}}{2}} \quad \text{by (7.12), (7.13), and (7.14)}$$

Practice 5

Solve

$$\left| \frac{1}{\log_{\frac{1}{2}} x} + 2 \right| > \frac{3}{2}$$

(Ref 2001 China)

First, by the definition of logarithm function, x must be positive. Then, $\log_{\frac{1}{2}} x \neq 0 \implies x \neq 1$. Hence, the basic requirement for x is

$$x \in (0, 1) \cup (1, +\infty) \quad (7.15)$$

Next, do casework.

1) If $\frac{1}{\log_{\frac{1}{2}} x} + 2 > \frac{3}{2}$, then $\frac{1}{\log_{\frac{1}{2}} x} > -\frac{1}{2}$

   i) If $\log_{\frac{1}{2}} x > 0$ or
$$x < 1, \quad (7.16)$$
   this inequality always hold.

Chapter 7: Solutions

ii) If $\log_{\frac{1}{2}} x < 0$ or $x > 1$, we have

$$\log_{\frac{1}{2}} x < -2 \implies x > 4 \tag{7.17}$$

2) If $\frac{1}{\log_{\frac{1}{2}} x} + 2 < -\frac{3}{2}$, then

$$\frac{1}{\log_{\frac{1}{2}} x} < -\frac{7}{2} \implies \log_{\frac{1}{2}} x > -\frac{2}{7} \implies x < 2^{\frac{2}{7}} \tag{7.18}$$

Taking (7.15), (7.16), (7.17), and (7.18) into consideration, we find the final answer is

$$(0, 1) \cup (1, 2^{\frac{2}{7}}) \cup (4, +\infty)$$

Practice 6

If the minimal and maximum values of function

$$f(x) = -\frac{1}{2}x^2 + \frac{13}{2}$$

in the domain $[a, b]$ are $2a$ and $2b$, respectively, determine the values of a and b.

(Ref 2000 China)

The absolute maximum value of $f(x)$ can be achieved as $\frac{13}{2}$ when $x = 0$. Therefore, maximum value $2b$ can only be one of $f(a)$, $f(b)$, or $f(0)$.

Correspondingly, minimal value $2a$ can be either $f(a)$ or $f(b)$.

1. If $a \leq 0 < b$, then maximum value $2b = \frac{13}{2} \implies b = \frac{13}{4}$. It follows that minimal value $2a = f(a)$ or $f(b)$.
    i) If $2a = f(b)$, then $2a = f(\frac{13}{4}) = \frac{39}{32} \implies a = \frac{39}{64} > 0$. This contradicts to the assumption $a \leq 0 < b$.
    ii) If $2a = f(a)$, then $2a = -\frac{a^2}{2} + \frac{13}{4} \implies a = -2 - \sqrt{17}$.
    Hence $\left[-2 - \sqrt{17}, \frac{13}{4}\right]$ is one possible solution.

Chapter 7: Solutions

2. If $a \leq b \leq 0$, then $f(x)$ monotonically increase in range $[a, b]$. Therefore we must have
$$\begin{cases} 2a = f(a) \\ 2b = f(b) \end{cases} \Longrightarrow \begin{cases} 2a = -\frac{a^2}{2} + \frac{13}{2} \\ 2b = -\frac{b^2}{2} + \frac{13}{2} \end{cases}$$
This means a and b are two roots of equation
$$\frac{x^2}{2} + 2x - \frac{13}{2} = 0$$
By Vieta's theorem, $a \cdot b = -13$ which means they have different signs. This contradicts to the assumption $a < b < 0$.

3. If $0 \leq a < b$, then $f(x)$ monotonically decreases in range $[a, b]$. Hence
$$\begin{cases} 2a = f(b) \\ 2b = f(a) \end{cases} \Longrightarrow \begin{cases} 2a = -\frac{b^2}{2} + \frac{13}{2} \\ 2b = -\frac{a^2}{2} + \frac{13}{2} \end{cases} \Longrightarrow \begin{cases} a = 1 \\ b = 3 \end{cases}$$

Therefore, in conclusion, the answer is $\left[-2 - \sqrt{17}, \frac{13}{4}\right]$ or $[1, 3]$.

Practice 7

Find the function $f(x)$ such that $f(0) = 1$, $f(\frac{\pi}{2}) = 2$, and for any $x, y \in \mathbb{R}$,
$$f(x+y) + f(x-y) = 2f(x)\cos y$$

Setting $x = 0, y = \alpha$:
$$f(\alpha) + f(-\alpha) = 2f(0)\cos\alpha = 2\cos\alpha \qquad (7.19)$$

Setting $x = \frac{\pi}{2} + \alpha, y = \frac{\pi}{2}$:
$$f(\pi + \alpha) + f(\alpha) = 0 \qquad (7.20)$$

Setting $x = \frac{\pi}{2}, y = \frac{\pi}{2} + \alpha$:

$$f(\pi + \alpha) + f(-\alpha) = -2f(\frac{\pi}{2})\sin\alpha = -4\sin\alpha \qquad (7.21)$$

Now, $(7.19) + (7.20) - (7.21)$:

$$2f(\alpha) = 2\cos\alpha + 4\sin\alpha \implies f(\alpha) = \cos\alpha + 2\sin\alpha$$

$$\therefore \quad f(x) = \cos x + 2\sin x$$

Practice 8

Let the domain of function $f(n)$ be $\mathbb{N}$, $f(1) = 1$, and for any integer $n \geq 2$,
$$f(n) = f(n-1) + 2^{n-1}$$

Determine $f(n)$.

Setting $n = 2, 3, \cdots$, respectively:
$$f(2) = f(1) + 2^1$$
$$f(3) = f(2) + 2^2$$
$$\cdots$$
$$f(n) = f(n-1) + 2^{n-1}$$

Adding them together gives
$$f(n) = f(1) + 2^1 + 2^2 + \cdots + 2^{n-1} = 1 + 2^1 + 2^2 + \cdots + 2^{n-1} = \boxed{2^n - 1}$$

Practice 9

Let the domain of function $f(n)$ be $\mathbb{N}$, $f(1) = 1$, and for any $m, n \in \mathbb{N}$,
$$f(m + n) = f(m) + f(n) + mn$$

Determine $f(n)$.

Let $m = 1$, then $f(n+1) = f(1) + f(n) + n = f(n) + (n+1)$.

Therefore
$$f(2) = f(1) + 2$$
$$f(3) = f(2) + 3$$
$$\cdots$$
$$f(n) = f(n-1) + n$$

Adding these relations gives:

$$f(n) = f(1) + 2 + 3 + \cdots + n = 1 + 2 + 3 + \cdots + n = \boxed{\frac{n(n+1)}{2}}$$

Practice 10

Let $f(x) : \mathbb{R} \to \mathbb{R}$. For any $x \in \mathbb{R}$, it always hold that $f(x+3) \leq f(x) + 3$ and $f(x+2) \geq f(x) + 2$. Define $g(x) = f(x) - x$.

i) Prove that $g(x)$ is periodical.

ii) If $f(998) = 1002$, find the value of $f(2000)$.

(Ref 2000 Beijing)

Let's start by trying a few numbers.

$$\begin{aligned} g(x+2) &= f(x+2) - (x+2) \\ &\geq f(x) + 2 - (x+2) \\ &= f(x) - x \end{aligned} \quad (7.22)$$

$$\begin{aligned} g(x+3) &= f(x+3) - (x+3) \\ &\leq f(x) + 3 - (x+3) \\ &= f(x) - x \end{aligned} \quad (7.23)$$

$$\begin{aligned}
g(x+4) &= g((x+2)+2) \\
&\geq f(x+2) - (x+2) && \text{by \textit{(7.22)}} \\
&\geq f(x) + 2 - (x+2) && \text{by given condition} \\
&= f(x) - x && (7.24)
\end{aligned}$$

$$\begin{aligned}
g(x+6) &= g((x+2)+4) \\
&\geq f(x+2) - (x+2) && \text{by \textit{(7.24)}} \\
&\geq f(x) + 2 - (x+2) && \text{by given condition} \\
&= f(x) - x && (7.25)
\end{aligned}$$

$$\begin{aligned}
g(x+6) &= g((x+3)+3) \\
&\leq f(x+3) - (x+3) && \text{by \textit{(7.23)}} \\
&\leq f(x) + 3 - (x+3) && \text{by given condition} \\
&= f(x) - x && (7.26)
\end{aligned}$$

By *(7.25)* and *(7.26)*, we have

$$f(x) - x \leq g(x+6) \leq f(x) - x \implies g(x+6) = f(x) - x$$

By definition, $f(x) - x = g(x)$. Hence, $g(x+6) = g(x)$ which means function $g(x)$ is periodic.

It follows that

$$\begin{aligned}
f(2000) &= g(2000) + 2000 \\
&= g(998 + 167 \times 6) + 2000 \\
&= g(998) + 2000 \\
&= f(998) - 998 + 2000 \\
&= 1002 - 998 + 2000 \\
&= \boxed{2004}
\end{aligned}$$

Chapter 7: Solutions

Practice 11

Find all functions $f : \mathbb{Q} \to \mathbb{Q}$ such that the Cauchy equation
$$f(x+y) = f(x) + f(y)$$
holds for all $x, q \in \mathbb{Q}$.

Taking $x = y = 0 \implies f(0) = f(0) + f(0) \implies f(0) = 0$.

Let's prove $f(kx) = kf(x), k \in \mathbb{N}, x \in \mathbb{Q}$ by induction. This is true for $k = 1$ Assuming it is true for k, then for $k+1$:
$$f((k+1)x) = f(kx+x) = f(kx) + f(x) = kf(x) + f(x) = (k+1)f(x)$$

Taking $y = -x$, we have $0 = f(x) = f(x+(-x)) = f(x) + f(-x)$. This implies $f(-x) = -f(x)$. Therefore $f(-kx) = -kf(x)$ holds.

Therefore, $f(kx) = kf(x)$ holds for all $k \in \mathbb{Z}, x \in \mathbb{Q}$.

Taking $x = 1/k$, then $f(1) = f(k(1/k)) = kf(1/k) \implies f(1/k) = (1/k)f(1)$.

For $m \in \mathbb{Z}$ and $n \in \mathbb{N}$, $f(m/n) = mf(1/n) = (m/n)f(1)$.

Hence $f(x) = cx$ where $c = f(1)$.

Practice 12

Solve the following system in integers:
$$\begin{cases} x_1 + x_2 + \cdots + x_n = n \\ x_1^2 + x_2^2 + \cdots + x_n^2 = n \\ \cdots \\ x_1^n + x_2^n + \cdots + x_n^n = n \end{cases}$$

Chapter 7: Solutions

Investigate the function

$$f(x) = (x-x_1)(x-x_2)\cdots(x-x_n) = \sum_{k=0}^{n} a_k x^k$$

where a_k, $(k = 0, 1, \cdots, n)$, are some constants.

Obviously, $f(x_1) = f(x_2) = \cdots = f(x_n) = 0$. Therefore

$$0 = \sum_{k=1}^{n} f(x_k) = \sum_{j=0}^{n} \left(a_j \sum_{k=1}^{n} x_k \right) = n \sum_{j=0}^{n} a_j$$

Hence we have

$$f(1) = \sum_{j=0}^{n} a_j = 0 \implies 1 \text{ is a root}$$

Or one of x_i equals 1.

Repeating this process can lead to the conclusion that all x_i equal 1.

Practice 13

Let x, y, and z be all in $(0, 1)$. Prove

$$x(1-y) + y(1-z) + z(1-x) < 1$$

Construct a function

$$\begin{aligned} f(x) &= 1 - [x(1-y) + y(1-z) + z(1-x)] \\ &= (y+z-1)x + (yz + 1 - y - z) \end{aligned}$$

The to-be-claimed conclusion is equivalent to showing $f(x) > 0$.

Because $f(x)$ is a 1st degree polynomial whose graph is a straight line, this above conclusion is equivalent to showing $f(0) > 0$ and $f(1) > 0$. This is because if the two points at both boundaries are above the x−axis, the whole line segment within this domain will be above the x−axis.

Because $0 < y, z < 1$, we find
$$f(0) = yz + 1 - y - z = (y-1)(z-1) > 0$$
$$f(1) = (z + y - 1) + (yz + 1 - y - z) = yz > 0$$

Therefore the conclusion holds.

Practice 14

Show that
$$\frac{(x+a)(x+b)}{(c-a)(c-b)} + \frac{(x+b)(x+c)}{(a-b)(a-c)} + \frac{(x+c)(x+a)}{(b-c)(b-a)} = 1$$

Construct function
$$f(x) = \frac{(x+a)(x+b)}{(c-a)(c-b)} + \frac{(x+b)(x+c)}{(a-b)(a-c)} + \frac{(x+c)(x+a)}{(b-c)(b-a)} - 1$$

The conclusion is equivalent to showing $f(x) = 0$ is an identity. Because $f(x)$ is a quadratic polynomial with respect to x, therefore it is sufficient to show that there exist three distinct zero points.

$$\begin{aligned} f(-a) &= \frac{(-a+a)(-a+b)}{(c-a)(c-b)} + \frac{(-a+b)(-a+c)}{(a-b)(a-c)} + \frac{(-a+c)(-a+a)}{(b-c)(b-a)} - 1 \\ &= \frac{(b-a)(c-a)}{(a-b)(a-c)} - 1 \\ &= 0 \end{aligned}$$

By symmetry, we can assert $f(-b) = f(-c) = 0$ must hold too. It is clear that a, b, and c are distinct because otherwise the given expression is not defined. Therefore, we can conclude $f(x) = 0$ always holds.

www.ingramcontent.com/pod-product-compliance
Lightning Source LLC
Chambersburg PA
CBHW071442180526
45170CB00001B/421